Computer/Telecom Integration

McGraw-Hill Series on Computer Communications

ISBN	AUTHOR	TITLE
0-07-003484-2	Ball	*Cost-Efficient Network Management*
0-07-005075-9	Berson	*APPC*
0-07-005076-7	Berson	*Client/Server Architecture*
0-07-005558-0	Black	*Frame Relay Networks*
0-07-005554-8	Black	*Network Management Standards*
0-07-005553-X	Black	*TCP/IP and Related Protocols*
0-07-005552-1	Black	*The V Series Recommendations*
0-07-005546-7	Black	*The X Series Recommendations*
0-07-010516-2	Cerutti	*Distributed Computing Environments*
0-07-016196-8	Dayton	*Multi-Vendor Networks*
0-07-016733-8	Dewire	*Application Development for Distributed Systems*
0-07-016732-X	Dewire	*Client/Server Computing*
0-07-019022-4	Edmunds	*SAA/LU6.2*
0-07-020346-6	Feit	*TCP/IP*
0-07-033754-3	Hebrawi	*OSI Upper Layer Standards and Practices*
0-07-028039-8	Heldman	*Future Telecommunications*
0-07-028030-4	Heldman	*Global Telecommunications*
0-07-032385-2	Jain	*Open Systems Interconnection*
0-07-034247-4	Kessler	*ISDN, 2/e*
0-07-034243-1	Kessler	*Metropolitan Area Networks*
0-07-035134-1	Knightson	*OSI Protocol Performance Testing: IS 9646 Explained*
0-07-042586-8	Minoli	*1st, 2nd, and Next Generation LANs*
0-07-046455-3	Naugle	*Local Area Networking*
0-07-046461-8	Naugle	*Network Protocol Handbook*
0-07-046320-4	Nemzow	*The Ethernet Management Guide, 2/e*
0-07-046322-0	Nemzow	*FDDI Networking*
0-07-046321-2	Nemzow	*The Token-Ring Management Guide*
0-07-051104-7	Radicatti	*Electronic Mail*
0-07-054418-2	Sackett	*IBM's Token-Ring Networking Handbook*
0-07-060360-X	Spohn	*Data Network Design*
0-07-063636-2	Terplan	*Effective Management of Local Area Networks*
0-07-042588-4	Minoli	*Imaging in Corporate Environments*
0-07-035968-7	Kureso	*Broadband Communications*
0-07-049663-3	Peterson	*TCP/IP Networking: A Guide to the IBM Environment*
0-07-038089-9	Lee	*Mobile Cellular Telecommunications Systems, 2/e*
0-07-060362-6	McDysan/Spohn	*ATM: Theory & Applications*
0-07-042591-4	Minoli/Vitella	*ATM & Cell Relay Service for Corporate Environments*
0-07-034232-6	Kerr	*Digital Cellular Telecommunications Systems*
0-07-049183-6	Pecar, O'Connor, & Garbin	*The M-H Telecommunications Factbook*

Computer/Telecom Integration

The SCAI Solution

Arkady Grinberg

McGraw-Hill, Inc.

New York San Francisco Washington, D.C. Auckland Bogotá
Caracas Lisbon London Madrid Mexico City Milan
Montreal New Delhi San Juan Singapore
Sydney Tokyo Toronto

Library of Congress Cataloging-in-Publication Data

Grinberg, Arkady.
Computer / telecom integration : the SCAI solution / Arkady Grinberg.
p. cm. — (McGraw-Hill series on computer communications)
Includes bibliographical references and index.
ISBN 0-07-024842-7 :
1. Computer interfaces. 2. Telecommunication systems.
3. Integrated services digital networks. I. Title. II. Series.
TK7887.5.G68 1995
004.6'2—dc20 94-35052
CIP

1 2 3 4 5 6 7 8 9 0 DOC/DOC 9 0 9 8 7 6 5 4

ISBN 0-07-024842-7

The sponsoring editor for this book was Marjorie Spencer and the production supervisor was Donald F. Schmidt. This book was set in Century Schoolbook by North Market Street Graphics.

Printed and bound by R. R. Donnelley & Sons Company.

To my mother, wife, and daughter.

Contents

Preface

Technical books are written to inform and to share ideas. When I thought of writing a book on SCAI, it was critically important for me to decide how to present the subject, and there are usually many alternatives to choose from. The initial choice affects the book's structure and the type of information presented because different goals are satisfied by different means.

SCAI, being a standard, can be discussed as the standard only, that is, a specific standardized solution for CTI. At the same time, the SCAI standard is created for the generic computer/telecommunications integration (CTI) market, which is emerging and expanding. In this book I attempted to cover two general directions: a presentation of the idea of integration and a specific, relatively narrow, initial approach to it. If a reader develops an insight into the opportunities, issues, and directions of CTI, I would be satisfied. I also hope that there is sufficient information contained in this book that can be used for practical purposes.

The process of writing a technical book is complex. Sometimes, difficulties appear when they are not expected. Help from colleagues was important and well appreciated by the author. There are many people whom I would like to thank for help in certain technical matters and in reviewing the book. Here I want to mention the names of several people whose help was critical: Dick Khan, Dan Minoli, Paul Roder of Bellcore who commented on the process and specific issues; Pete Russo, Greg Ratta, Jim Garrahan of Bellcore, who reviewed the book and helped improve its clarity; Joeff Gilbert and Mike Ryan of IBM, Peter Storrie of NEC, and Denise Brittin of Intecom, who responded to my requests for information. I also want to acknowledge my employer, Bellcore. While the project was not sponsored by Bellcore, it was supported in spirit.

Arkady Grinberg

Chapter

1

Introduction

The information age implies the ability to access information of all types and use this information in many aspects of our lives. The growth in the number of alternative ways of accessing information and of the information content is startling. It seems that new services and technologies appear overnight. In business and home environments, people learn to rely more and more on automated informational tools—computers, communications systems, electronic gadgets.

Various branches of the information technology tree are growing more or less independently from each other. When the development of some of these branches reaches certain milestones, the industry usually steps back and begins to contemplate the idea of merging some of them. This is currently the case with telecommunications and computing technologies. A strong need to provide services common across telecommunications and computing environments generated the idea of *computer / telecom* integration (CTI).* This is a generic term referring to a variety of technological solutions and directions in the area of the integration of two environments. It addresses a trend toward combining resources of two types of networks, telecommunications and data processing (computing), into a single environment to provide more sophisticated services to users.

1.1 What Is SCAI?

There are a number of approaches to CTI that are different in scope and implementation. Manufacturers of telecom and computer equipment

* The word *telecom* is short for *telecommunications,* which is typically used in this context.

developed CTI solutions based on their products, and each solution corresponds to the vision that each manufacturer has regarding CTI.

One CTI solution developed by the industry is the American National Standard. This standard was defined by Committee T1 and is titled Switch-Computer Applications Interface (SCAI—pronounced "sky"). The SCAI standard addresses a number of applications (services) and provides the foundation for the evolution of the technology to support a greater number of services in the future.

The basic idea underlying the standard is very simple: switches inform computers of certain stages of call processing and computers take certain actions based on this information. Also, computers can direct switches to perform certain actions. The SCAI link shown in Fig. 1.1 connects a switch and a computer, or, using a more generic view, connects telecommunications and computing environments. The exchange of information that takes place over the SCAI link between switches and computers results in services that can be characterized as integrated computing/telecom services.

Figure 1.1 shows a basic SCAI configuration. This configuration supports a variety of call management applications (services), including inbound telemarketing, outbound telemarketing, and a customer ser-

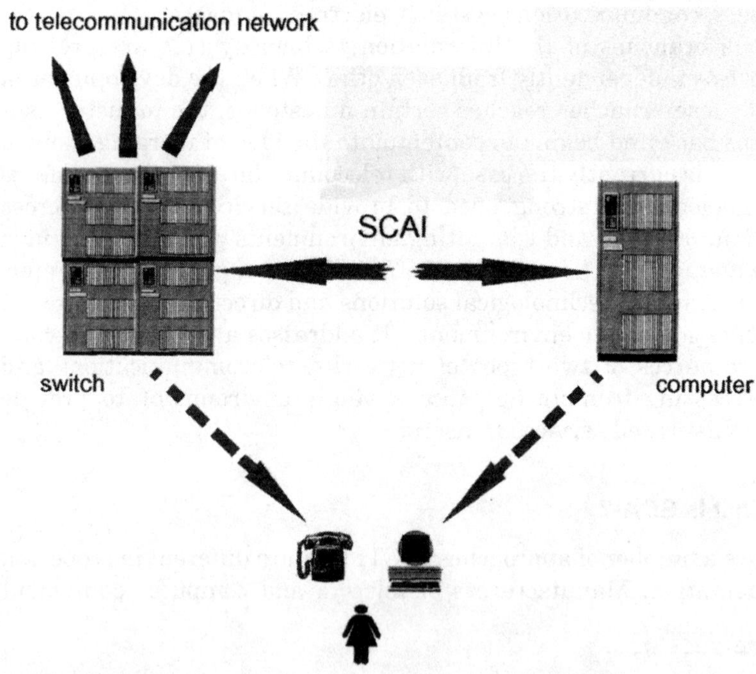

Figure 1.1 SCAI.

vice center. The exchange of information over the SCAI link benefits an agent (someone using the telephone and the computer terminal as illustrated in Fig. 1.1) by simultaneously delivering to this agent a voice call and a data screen (information about the caller). The SCAI technology used in such an arrangement provides benefits to its users and to people who establish contact with organizations using this technology. Deployment of SCAI-based technology saves resources, helps to provide better, more personalized services to customers, creates a better working environment for agents, and provides a solid platform for a possible evolution to new CTI services.

An example of an application that can be supported by this configuration is inbound telemarketing. In this application, telephone calls arrive from the telecommunication network to the switch with the purpose of reaching an agent to purchase something (catalog home shopping). When a particular call arrives at the switch, the computer is informed about it over the SCAI link. The computer directs the call to a particular agent and simultaneously delivers information regarding the calling customer to the same agent's computer terminal. The agent answers the call and can immediately serve the caller (customer) because information about the customer's name, address, preferences, and so on, is already available.

SCAI, being a standard, also provides the added benefit of open opportunities for users who install SCAI-compatible equipment. This is a benefit that any standard-based solution affords to its users: the option of buying equipment from multiple vendors, which alleviates the possibility of being stuck with a manufacturer who somehow does not adequately respond to the customer's needs.

The SCAI standard (T1.626) contains the following functions:

- Basic calls (making, clearing, answering calls)
- Call Transfer (transferring calls from one device to another)
- Call Conferencing (conferencing several participants into a single call)
- Event Reporting (reporting specified events occurring in a switch to a computer)
- Call Hold/Retrieve (placing participants in calls on hold and later retrieving them)
- Routing (obtaining routing information for a call from a computer)
- Monitoring (establishing controls for specific devices)
- Security (securing communication between a switch and a computer)

The standard describes the functional architecture, protocol architecture, SCAI objects, modeling of calls and other dynamic objects, and a specific protocol that has been defined using ASN.1. The SCAI standard published by ANSI in the beginning of 1993 is not the final product; it has certain limitations and these limitations will be addressed in future releases of the standard. However, the foundation of the SCAI standard—its architecture, basic objects, and approaches to creating and describing functions—will not change. Therefore, users of SCAI equipment can rely on its evolution together with the evolution of the standard.

1.2 Who Should Read This Book?

This book is based on the approved American standard and does not attempt to copy the standard and present its every element. Instead, it focuses on the following goals:

- To present and explain a new standard and how it can be used. In addition, while the standard is self-contained, experience shows that there are usually aspects of any standard that need more detailed explanation than the standard text can afford.
- To discuss the general issue of computer/telecom integration, the reasons for integration, and the opportunities that this technology presents.
- To show how abstract concepts are being applied to new development. Specifically, the abstract concepts of OSI and object-oriented design played a significant role in the SCAI standard development. This book addresses the issue of the practical application of these concepts.
- To present and explain the standards process, specifically the process of standards development in Committee T1.

All these goals are important and probably deserve a book focusing on each of them. This book, though, attempts to combine the presentation of a new technical standard based on the new technology with explanations of how and why certain things occurred. This should help people with diverse interests and, at the same time, makes the book a reference for someone closely involved with CTI. Whether any of the goals have been realized is for the reader to decide.

Several types of readers should benefit from this book:

1. A designer of a specific SCAI product who needs to know more about the SCAI standard than the text of the standard presents. Sometimes, the supposedly clear definitions are not so clear to people not involved with their development.

2. A business manager responsible for customer service, security, or some other client-oriented task in an organization who needs to learn about the technology that helps the bottom line of the business.
3. A technologist who wants to stay abreast of new technologies. In the same category are people of other professions who are also interested in learning about new technologies: marketing personnel, business managers, engineers in fields other than computers, college professors.
4. A user, someone who uses the technology in business. Managers of telecommunications and data processing departments and planning and engineering personnel of the same departments need to understand the technology available to them.
5. A student or a young engineer who needs to learn about the "rules" of developing new products and new communications protocols. This book identifies some issues associated with the protocol design and implementation that are of importance to anybody who attempts to define high-level protocols to support new applications.
6. An engineer who needs to learn about the applicability of abstract concepts such as OSI to a specific task.
7. Anyone who needs to learn the process of developing standards. This process can benefit its participants, but only if the written and unwritten rules are known.

1.3 The World of Standards

The world of telecommunications and computing standards is complex and varied. Many different standards committees currently exist. Many of them are important to the interests of the industry and the public. There is a certain language used by the standards community, and this is not different from any other community of human beings. Without the knowledge of this language it is difficult to understand what is going on. What makes this language, in addition to the special technical jargon, can be described as names of various committees (T1S1, X3S3, WP 11/4-2), titles of standards (Q.921, I.310, T1.607), and special words pertaining to the process (*Resolution 1, interim meeting*). The standards' titles and the names of the committees producing these standards are used in the press and by the interested public quite often. Many times, people who use them do not know any details of the standard or the committee that developed it, but do know what the particular standard is expected to do.

Two types of standards are referenced in this book: ANSI and ITU-T. ANSI is an American standards body and ITU-T is international. The names of ANSI standards are prefixed by "T1" and have the general form of "T1.LMN," where the letters *L, M,* and *N* signify a number. The names of ITU-T standards begin with a letter (each letter identifies a particular Study Group that developed a standard (e.g., *Q* identifies Study Group 11) and have the general form "W.LMN" (Q.971). Some people are familiar with the international telecommunications standards committee CCITT. In 1993 it was renamed ITU-TSS, which stands for International Telecommunication Union-Telecommunication Standardization Sector, and is addressed in most instances as ITU-T.

In many cases, standards for the same interfaces and functions are produced by more than one committee. Usually, these are regional (such as T1) and international (such as ISO) committees. Sometimes, two international committees produce the same or similar standard (ISO, ITU-T). And sometimes multiple regional committees produce the same or similar standards (such as T1 and TIA). In this book a number of standards are referenced, especially in the chapters describing transport mechanisms and OSI concepts. Most of the mentioned standards have equivalent American and internationally produced functionality hidden behind different names (e.g., T1.607 and Q.931 standards* address the functionality of the ISDN layer 3 signaling protocol and are very much the same), or equivalent ITU-T and ISO-produced functionality also hidden behind different names (for example, ISO 8649 and ITU-T X.217 address ACSE and are almost exactly the same). In these cases, only one standard is referenced, sometimes T1 and other times ITU-T (the same is true for ISO versus ITU-T). Some older versions of standards are mentioned as CCITT standards because a new ITU-T standard would be different. The reader should be aware of this situation and can check the references at the end of this book if there is a need to correlate standards produced by different organizations.

1.4 How to Read This Book

The book presents information in the sequence that I believe is the most beneficial to a reader who does not know much about this particular technology, the specific standards mentioned in the book, or the standards process. However, it is clear that the backgrounds of different readers are not the same and they would have different degrees of familiarity with the concepts described in the book. I have attempted

---

* In the ITU-T case, these are called *Recommendations,* but they are standards.

to structure each chapter in a way that allows it to be read with sufficient benefit without reading other chapters. This will not work in all cases, but should work in some. Forward and backward references are used to direct a reader to another chapter that gives more details on a subject.

Chapters 2 and 3 give a historical perspective on the issue of computer/telecom integration and outline the reasons for the need to merge the two environments, telecommunications and data processing.

Chapter 4 is a general tutorial on the concepts used in the development of the SCAI standard. The concepts are OSI and associated standards, ASN.1, and object-oriented methodology. A reader can skip this chapter or any part of it if he or she is sufficiently familiar with these concepts.

Chapter 5 presents the details of the SCAI standard. Each part of this chapter addresses specific SCAI issues—architecture, protocol, conformance, and so on. It is most beneficial to read the entire chapter, but, if a reader is familiar with a certain aspect of SCAI, that particular part can be skipped. For example, if a reader is familiar with the SCAI architectural approach and wants to learn more about the call processing aspects of SCAI, the functionality section can be studied.

Chapter 6 discusses possible applications (services) that SCAI can support. This chapter is the key for those who want to understand the applicability of this technology to their particular situations. This chapter addresses the applications that the SCAI standard is designed to support and gives examples of applications that it can support with some enhancements.

Chapter 7 addresses the extremely important issue of transport mechanisms for SCAI. Although transport mechanisms are not part of the standard, they are as important for the provision of services as the standard itself. Without the transport, the SCAI information cannot get to its destination. This section is particularly important to those who are deploying SCAI equipment.

Chapter 8 outlines a process to follow when applying the SCAI standard to a specific implementation. It discusses the alternatives that exist in using SCAI. It also addresses the issue of how to decide whether the SCAI standard is applicable to specific customer environments. Therefore, this chapter would be of great use to implementors, users, and those thinking of deploying SCAI or, actually, any CTI product.

Chapter 9 discusses the applicability of IVRs in the SCAI environment.

Chapter 10 focuses on the future development of SCAI.

Chapter 11 gives examples of three particular implementations of the CTI technology. Equipment from IBM, Intecom, and NEC is discussed to show the reader how nonstandard solutions address the same issues as the SCAI standard. This chapter would be of interest to any reader because the equipment discussed can be seen, touched, and even bought and installed. Although the book discusses the standard solution, the world has not stood still waiting for it and examples of earlier implementations should be of interest to the reader.

Chapter 12 presents the details of the standards process that the Committee T1 follows. The structure of T1, its processes, and suggestions on participation in any of the T1 organizations are outlined.

Some portions of the T1.626 standard are included in the appendix, specifically section 10, which defines the protocol to perform the SCAI functions. This section is included to aid the reader who does not possess the SCAI standard in understanding this standard.

HAPPY READING!

Chapter

2

Computer/Telecom Integration

This book is concerned with the subject of the integration of two *technologies,* or *activities,* or *concepts.* These two technologies started and developed separately, and they were used for different reasons in different situations. Years ago, when telephone networks were already an important element of people's lives and computers were designed to perform calculations, no issue of integration existed. Since that time many things have changed. While, in general, the telephone networks are still primarily used to make calls (voice calls), computers have been transformed from large calculators to widely used tools. In the process, the computing machines—computers—transformed society. These tools are now used not only by professionals with technical degrees, but by everybody, sometimes even without an awareness of them.

While the phenomenon of the computer's entry into everyday life is not new and is certainly not the subject of this book, it is still the primary reason for us even to contemplate the subject of computer/telecom integration. Before we attempt to identify the practical drivers for such integration, we need to investigate what separates and unites telecommunications and computer networks.

2.1 Telephone Networks

Telephones and, later, telephone networks were initially designed for a specific visible task: to allow people located in different places to communicate with each other. At first, some observers were sure that the new technology would be used only in special instances, for example, to relay business information from one office to another or to request emergency services. We all know how wrong those observers were. Many people cannot imagine their lives without telephones, and the

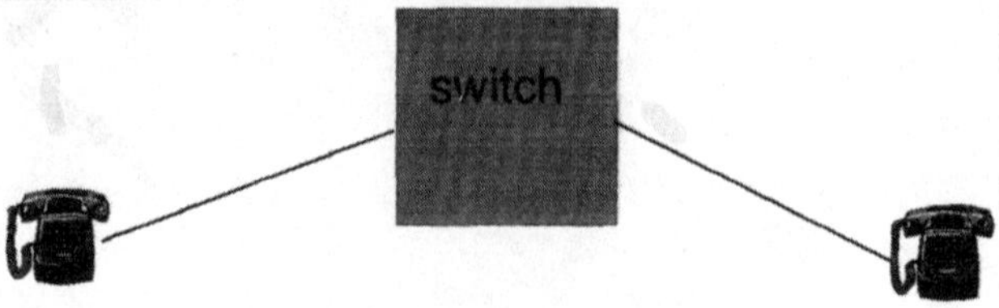

Figure 2.1 Switching of telephone circuits.

availability of telephone service is commonly used as an indication of the general health of a nation.* The telephone is truly one of the most remarkable consumer devices ever created. But in the beginning, telephone service was simple.

The first connections used for two people to communicate were set up by human operators on the verbal request from someone initiating a call. Then a telephone switch was created (see Fig. 2.1). A switch freed human operators to do other tasks and allowed connections to be established automatically.

The goal of any business is to make money, and so telephone service providers were interested in finding ways to (1) sign up more customers and (2) create new attractive services. The next step in the evolution of telephone service was linkage of individual switches together, or the creation of a telephone network (see Fig. 2.2).

The telephone network allowed connections to be established over larger distances and, therefore, allowed people located at great distances from each other to communicate. The first network was just a collection of individual switches linked together to allow a long distance communication to take place. With the growth of the number of switches and the distances between them, there was a need to find an optimum solution for networking, that is, to connect switches in the most beneficial fashion. The next step in the evolution was creation of a hierarchical network. In such a network, only specific switches are interconnected, each having its own cluster of switches that it "oversees." Figure 2.3 shows how the hierarchy can be accomplished.

With the growth of telephone networks, the networks' capabilities also grew and, therefore, services that could be provided by them expanded. The development of telephone networks was geared for a larger coverage area and better-quality connections until switches composing the networks became computerized. The first stored program control switches appeared in the 1960s. These were the switches equipped with computers that controlled their operations. Although

* I wonder why? Without telephones, people write letters; with telephones, people forget how to write.

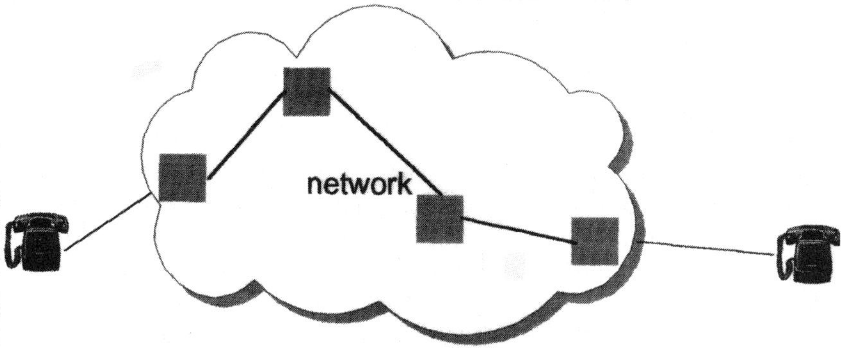

Figure 2.2 Telephone network.

the first computers in the switches were highly specialized, they were computers nevertheless. The change from electromechanical to computerized switches was nothing less than a revolution in telephone networking. From that moment on, the technological evolution supported not only better networking, but more opportunities in service creation. Services such as *hold/retrieve* began to appear, and then *call forwarding,* and more.

Relatively speaking, computers controlling switches were easy to program to perform these new types of services. As computers matured and faster and more powerful computers were used in the switches, more services, such as *call waiting,* were offered to the customers. With computers, the development of telephone services and telephone networks experienced a completely new dimension. Digitization of switches and transmission added impetus to the new dimension. The telephone networks' evolution has reached the state at which a large number of calls are being controlled, many services are available, and great potential exists for the creation of new services—all of it due to the computerization and digitization of existing well-developed networks.

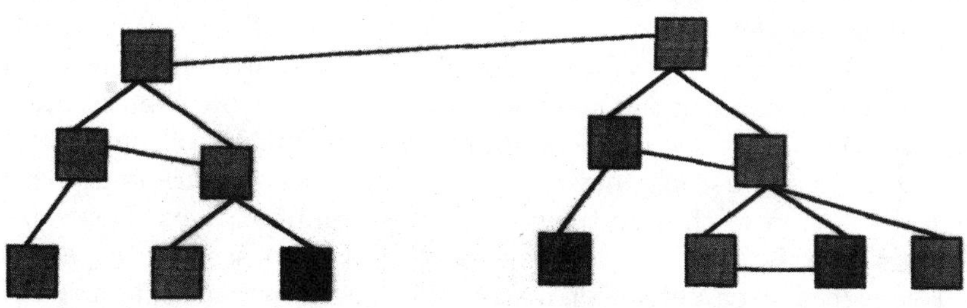

Figure 2.3 Hierarchical network.

So far, the discussion has been about the public networks and the switches composing them. Beginning in the 1970s, another family of switches appeared. These were private branch exchanges (PBXs). A PBX is a computerized switching system designed for use in private networks. Large corporations, governments, and universities buy them and create their own networks—private networks. These networks interconnect various locations and also connect to public networks for public communication. A PBX is basically the same switch as the one used in public networks. There is a difference in the services that a PBX and a public network switch provide (although Centrex services of public networks attempt to match the PBX services) and a difference in some performance requirements, but the intent of both types of switches is the same: to establish, manipulate, and release calls.

2.2 Computer Networks

Computers were designed to perform calculations and to retain information. Initial calculations were of a scientific and accounting nature only. As computers became more powerful, a greater variety of tasks was delegated to them. The original computers were very large and occupied a lot of space. They were installed in special rooms and were used by organizations that were able to acquire them to serve as powerful calculators. Since all decision making is based on calculations and comparison, computers were easily adapted to perform tasks other than straightforward scientific calculations.

With time, computers became smaller in size and more powerful. From the mainframes to the minicomputers and then to the first microprocessors, the evolution continued. More powerful machines could perform greater numbers of operations and, at the same time, were smaller in physical size than their predecessors. A combination of greater capabilities, smaller size, and, eventually, lesser price propelled the spread of computers into all aspects of modern life. Computers are now running cars and helping to serve customers, teaching students and controlling technological processes, performing translations and maintaining databases of information. Examples are numerous and any reader is familiar with many of them. What is important, though, is to see that the evolution of the computer brought a powerful calculator to the status of the most powerful tool people ever created. Today's information age is the result of the development of computers.

Computers are also organized into networks, but the variety of these networks is greater than in the case of telecommunications. There are all kinds of configurations created for different purposes: public packet networks, local area networks, metropolitan area networks, and so on. While the existence of telephone networks was based on the need of people to communicate, the existence of computer networks is due to the need for information sharing.

2.3 Integration

Our short historical detour brings us to the present time, which is characterized by highly developed telephone and computer networks, each serving specific needs of humanity. The obvious question is whether it is useful and possible to merge the two types of networks, or, as we say, two *environments.*

The state of development of computers, switches, and networking techniques is such that a high degree of integration of telecommunications and computing (data processing) environments is possible. The issues of information representation and transfer* are well advanced and would not be a barrier. Also, the understanding of networking issues, although not perfect, is sufficient to consider merging two types of networks initially designed for different tasks. Computers running various systems (switching and computing) are fast and have access to large banks of memory where information is stored. This allows the introduction of new software to support new services. In short, the state of technology is not a problem. What about the need?

There are two general groups of people in the telecom/computing world: those that provide services and those that receive services (users). These groups are different in their relation to the potential integration. The providers are interested in business opportunities, in better services to their customers, and in decreasing expenses. The users are interested in the technology that makes their lives easier, supports them in their endeavors (either their business or private interests), and does not cost too much.

With the spread of computers, users are significantly more sophisticated and are ready for new services, some of which they cannot even perceive. A lot of attention is being paid to marketing and developing products that someone will immediately buy, and this is definitely the correct approach. However, there is also a place for the technology to push the envelope and introduce new products that find better or more complete usage only after some time. There is a possibility for technology to provide a foundation on which totally new services are created later. The integration of computer and telecommunications

---

* By information representation and transfer, I mean all the software engineering techniques of representing information in the computer memory, information manipulation, and transfer between different applications; in short, the state of computing techniques.

environments, in my opinion, is an example of such a technological development.

But this integration would not rely solely on future usage; currently there are market signals that indicate that users would benefit from it right now. Users would have easier access to a greater number of existing services from their current devices (telephones and computer terminals). More of the complexity of dealing with specific networks would be hidden from them. Some of the tasks would be accomplished faster, allowing more efficient use of resources. In short, if computing and telecommunications services can be accessed the same way, the user benefits from it.

Providers of services comprise another group that can benefit from the integration discussed here. First, service providers are interested in finding ways to increase their customer bases. If services are easier to use, more customers will sign up. Second, a more sophisticated technology gives impetus to the development of new ideas and services, the realization of which is a source of new revenues. Third, one of the ways to decrease expenses is to minimize the costs associated with provisioning and administering networks, training personnel, and performing maintenance. If all of these activities can be made sufficiently similar such that they do not require two or more sets of completely different procedures, significant resources could be saved.

Our short analysis has provided us with the reasons for the high degree of integration. This, of course, is not revolutionary and has been recognized by a number of companies involved in the telecommunications and computing industry. Some of these companies developed equipment to start the process of integration, and some of them participated in the development of public solutions, that is, solutions which are sought in public forums. We will discuss both in the rest of this book.

Chapter

3

SCAI as an Integration Technology

The discussion of computer/telecom integration requires an understanding of the nature and scope of such integration. *Integration* is one of those words that seems to be used to cover ambiguity, similar to *service* and *architecture*. Webster's dictionary defines integration as an act of "making into a whole." So, what kind of integration should be pursued to respond to the needs discussed in the previous chapter?

There are already efforts underway to integrate certain aspects of telecommunications networks. For example, ISDN was developed to integrate voice and data access to networks. This was achieved by combining what used to be different network access technologies into a single access mechanism. Termination of the ISDN access lines carrying voice and data traffic by a network was also integrated. Another example is ATM, developed to integrate transmission of all kinds of information. This was achieved through the creation of new transmission techniques that allow the same handling of voice, data, video, and other forms of information. A single transmission medium is used in all cases. Both ISDN and ATM focus on technological issues, although they ultimately will impact network users.

SCAI begins with users. It proposes integration from the users' perspective. Users do not usually pay much attention to the details of the technology that serves them as long as that technology performs up to their expectations. They need easy access to many services that telecommunication and computer technologies provide. SCAI creates a single environment instead of the existing two (telecommunications and data processing). Users benefit from SCAI through greater productivity, better service, and a greater ease in using technology.

SCAI can be looked at as a tool that provides users with easy access to both environments. In other words, SCAI is a tool for creating a single environment. A single information environment, as we can call it, changes many notions that currently exist. It is usually the rule in a large company to have two distinct corporate departments to deal with telecommunications and data processing. Today, telephony people handle the former and MIS people handle the latter. The need is historical and has existed for a good reason. SCAI changes this very practical aspect of corporate life because, from the technological point of view, there should no longer be a need for two distinct approaches to the treatment of information. SCAI creates a single environment for users who can now develop a single approach to this new *information* environment.

There are three distinct components of the integration approach that is taken by SCAI. To achieve a meaningful degree of integration, all three components must be developed. These components are:

1. The abstraction of applications in the telecommunications environment
2. The abstraction of applications in the data processing environment
3. The link that merges the two environments

To understand the role of each component, we need to discuss in more detail what is actually going on in the two environments.

3.1 The Telecom Component

The first component is telecommunications-related. The telecommunications environment is characterized by calls that are set up, released, forwarded, and otherwise manipulated. The main entities in telecommunications networks (switches) perform the task of call processing. Call processing is manifested in software designed based on the algorithms that were developed to handle all kinds of calls. Every time a call is attempted by a user, software in a switch reacts to it according to the algorithm for this particular type of call. Any kind of integration between two environments must take into consideration how call processing is designed and behaves. Call processing is a major application in telecommunications networks. There should be a way by which entities outside of telecommunications networks can "see" the processes taking place within them. The call processing application can be abstracted into a sequence of meaningful states that are visible to the outside world. This abstraction is a call model.

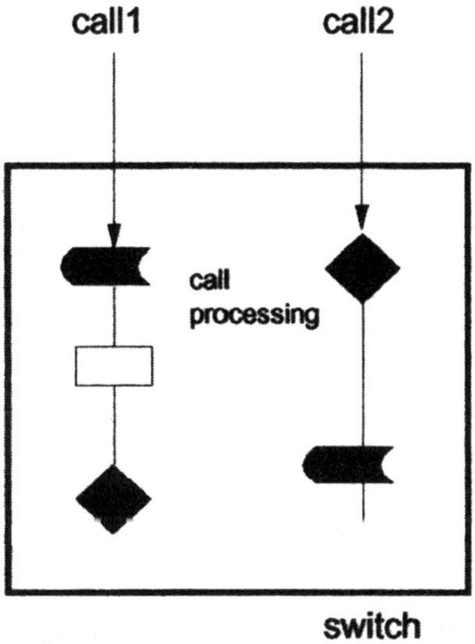

Figure 3.1 Call processing in a switch.

It represents the environment and, for call processing purposes,* it is the point of contact with abstractions (models) representing the data processing environment. At the same time the call model hides the details of call processing that are of no interest outside of the telecommunications environment.

Figure 3.1 shows a representation of call processing as an abstract call model. Two calls enter the switch and are handled according to the appropriate call processing algorithms. These two calls happen to go through different states because they are in different situations; for example, call 1 reaches its destination and call 2 does not (it failed). These two calls take different paths through the same call model.

The call model consists of call states corresponding to different activities in the telecommunications network. Let's discuss these activities

---

* The question can be asked whether call processing is the only application operating in the telecommunications network. The answer to this question largely depends on the definition of call processing. If call processing means only processing of voice calls, then there is definitely more than one application in the telecommunications network. Examples of others: data communications, video communications, etc. We can also state that all types of communication are not different in their goal, which is to set up, release, and manipulate "calls" of all sorts. If we take this view, we need to carefully examine if call processing for different types of calls is the same, and this would depend on the technology used for these various types of information exchanges. It is probably best to imply voice communications when we discuss call modeling as it applies to SCAI. Hence, the conclusion must be that when SCAI is extended to incorporate activities of telecommunications networks beyond voice calls, we can refer to these new activities as applications different from call processing.

in more detail. When someone makes a call, the network performs several tasks, the goal of which is the creation of a communications path to allow two parties to talk. When this goal is achieved, it is said that a call is established. Although there is no need to list all the very granular steps in the process of call establishment, it is useful to present the key decision points involved in call processing. Each decision point represents a different state in the call model, and these decision points are as follows:

1. When party A picks up a phone, a switch recognizes the call attempt and gets ready to receive digits.
2. Party A dials a number to reach party B, and a switch checks for the completeness of the dialed number.
3. A switch analyzes the dialed digits and finds an outgoing route to reach party B.
4. Party B is alerted to an incoming call through the application of a ring.
5. If party B answers, the call is established and the conversation between parties A and B takes place; if party B does not answer, party A hangs up and the network clears the call; if party B is busy, party A hangs up.*
6. When any party hangs up, a switch (actually, a network) releases all established connections and frees resources.

The abstractions of the telecommunications network applications—in our case, the call model—is the first component of the integration.

3.2 The Computing Component

The second component of the integration is data processing related. The data processing environment is characterized by the focus on information and a huge variety of application software that uses information for various purposes. Computers are becoming so widespread and being used for so many different reasons that the variety of software applications is tremendous and growing. Computers are gaining acceptance in more and more fields as their power is increasing and cost is decreasing. Many familiar places that did not have computers only a short time ago are changing as the information age takes hold on society. If yesterday computers were used by a few professionals,

* This discussion does not address the availability of services such as call waiting, call forwarding, etc. The point of the discussion is to show general steps for call establishment.

today they are used by almost everybody. It is becoming more of a rule to use computers as tools that perform complex tasks quickly and precisely. Many computer users do not know and do not need to know how the tasks are being performed. Most of the complexity is hidden from the users. New user-friendly interfaces are being designed that help increase the number of users and, in turn, the number of applications that computers support.

The variety of applications in computers and computer networks is much greater than the variety of different software elements in the telecommunications switches. This difference creates distinct problems and concerns. Some of these concerns are related to the programmers' job, which is to write software for a variety of environments using different operating systems running on different computer hardware. There is a need to allow the same program to use different resources that perform similar functions, and to allow various programs to use the same resources. There is also a need to run programs on many different computers without making a great many changes. As a response to these and other concerns and problems, the API concept has been created.

Since the time the term API became widely used in the industry, I have seen API defined as application programmer interface, or application program interface, or application programming interface. While any variation of the definition is acceptable, I do think that a single term is necessary. Since I prefer the term API to mean *application programming interface,* this is what it stands for in this book.

An API hides from applications the details of the inner working of some resources. It can be an API interfacing to a variety of operating systems, so the application is portable on many computers. Or it can be an API interfacing to a communication system for the exchange of information. In the SCAI case, the API would allow applications to be free of concerns about SCAI details regardless of the origin of a switch.

In Fig. 3.2, applications A1, A2, and A3, running in the same computer and supporting different services, have access to other computer resources through an API. An API performs some of the translations that are common to all applications. Without an API, each application would have to perform the same tasks of interfacing to needed resources.

SCAI incorporates the representation of call processing in the form of a call model. On the computer side, the data processing model is not considered. It would be either too complex, if behavior of many applications were modeled, or too trivial, if only the internals of computers (operating systems) were modeled. Instead of a computer model, the API should serve as the representation of the data processing world, in other words, of applications providing services to users. An API is an

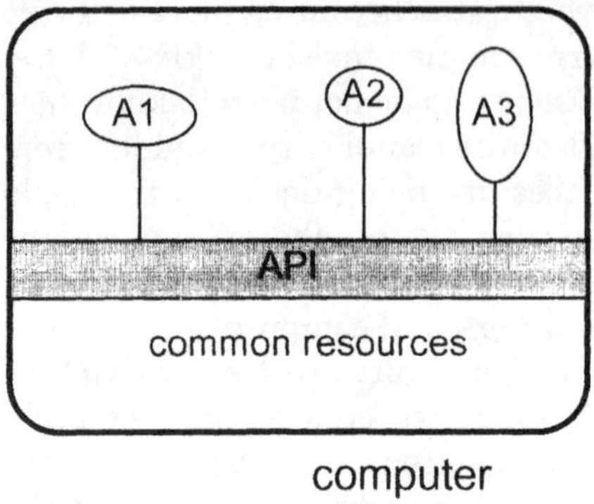

Figure 3.2 API.

abstraction of the data processing environment and is the second component of the integration.

3.3 The Integrated Link Component

The third component of the integration is the interface between the two environments. The definition of this interface requires an understanding of architectural issues and functions, and involves the design of the specific protocol. The interface must be in tune with the call processing of the telecommunications environment and with the developing API of the data processing environment. Of course, the latter is more difficult because the development of an API is just underway and many aspects of it are not clear.

SCAI ties the call processing in a switch with an API in a computer through a number of protocol elements (messages) that carry information between them. Because the API is currently not standardized, the SCAI standard is actually defined with the expectation that some form of an API exists, but it is not described in the standard.

Figure 3.3 represents a model of the integration of the two environments, telecommunications and data processing. In fact, it is now a single information environment created using SCAI as a link. Being a link between two distinct environments implies that SCAI, through its functionality, influences the development and evolution of both environments, especially data processing. This influence on computing will be based on new functions that SCAI would allow computers to access, specifically functions that are performed by telecommunications networks.

We talk about SCAI linking two environments, but the model of integration includes a link between a switch and a computer. The terminology we use is also based on the notions of a single switch and a single computer. So what is really implied by the model?

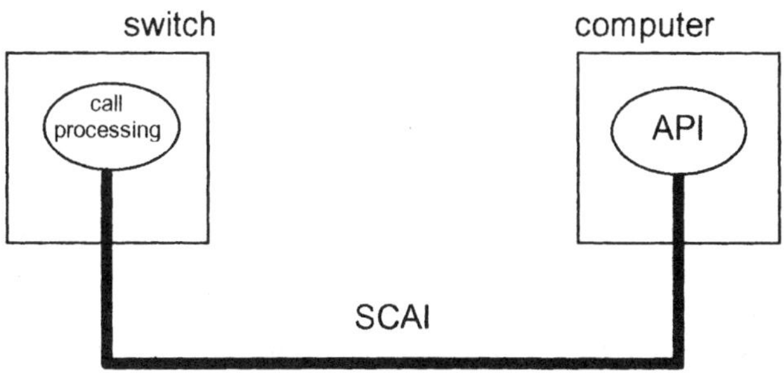

Figure 3.3 SCAI.

Obviously we mean more than a single network entity when the connection between a switch and a computer is shown. Telephone calls in the network are usually handled by more than one switch. Computing actions are often performed by more than one computer. We choose to represent an integration link (SCAI) as a link between single entities, but the meaning of it is a link between two distributed entities. In one case, the distributed entity handles calls (telecommunications environment), and in the other case, the distributed entity performs computing tasks (data processing environment).

Figure 3.4 shows the integrated information environment. A network of telecommunications entities constitutes the telecommunica-

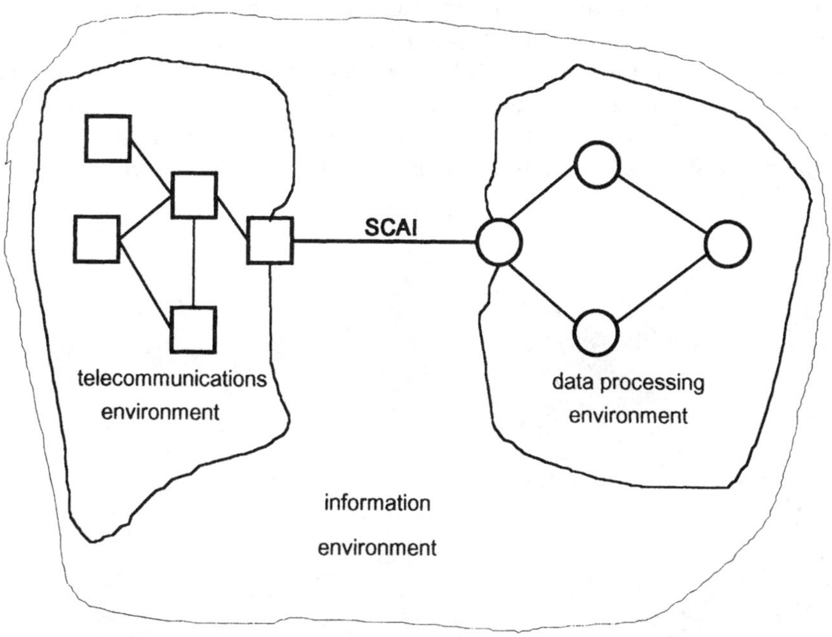

Figure 3.4 Distributed environments.

tions environment. These entities are not only switches, but all necessary components of telecommunications networks, for example, operation systems, databases, and so on. All of them together are addressed through a single switch connected to a SCAI link. When we talk about call processing and call modeling, we mean call processing in the network because there is no way of predicting how a network would route and manage the call. To SCAI, it seems like an activity occurring at the connecting switch, but in reality it may be different. SCAI would not be concerned with the details of the network (given that the network generally supports SCAI applications) and would not distinguish between an analog network, ISDN, or IN. SCAI expects only certain behavior represented by the call model.

On the other side of the SCAI link in Fig. 3.4 is a network of computers constituting the data processing environment. SCAI is connected to one single computer (gateway) and expects behavior as defined by SCAI models. Actually, SCAI interfaces with a gateway computer's API, which hides not only specifics of applications, but also specifics of a network of computers that may support an application. This network may be a LAN, MAN, or something else. An application running in a gateway computer may need an access to database servers for information; the applications may even be distributed among several computers, but all of these details are not visible to SCAI.

The question may be raised regarding multiple SCAI links between two environments and their relationship to the issue we are discussing. It has to be clear that, regardless of the number of SCAI links, the principle is the same: each link connects two environments and is a tool for creating a single information environment. Whether multiple links are to be used is not a modeling, but an engineering, issue. It is considered in Chap. 5 from the point of view of the SCAI architecture.

Chapter

4

Background Information

Before we can discuss the SCAI standard it is useful to review the concepts and existing standards that were used in the development of the SCAI standard. These concepts are OSI and object-oriented modeling, and the standards are ASN.1, ACSE, and ROSE. This chapter provides tutorial information that will be useful in understanding the SCAI standard. If more information is required, the reader should use the latest standards or any of the commercially available books that cover these subjects in more detail.

4.1 OSI

One concept was followed in the development of the SCAI standard, just as it was followed and is currently used in the development of many other standards in telecommunications. This concept is *Open System Interconnection (OSI)*. Interconnection of systems is necessary when there is a need to communicate. Communication requires protocols—sets of rules governing the exchange of information. An open system provides an open interface to its communications capabilities. Any other system can communicate with it, provided the two systems negotiate a specific protocol to be used. An open system concept, therefore, promotes interconnectivity, or networking, of systems through a set of agreed rules that establish an open environment. Communications systems (or software) designed using an OSI approach can be introduced in multivendor networks faster and more easily. So, what are the principles of OSI?

The first principle is *layering*. The functions that any communications system performs are subdivided into functional sets. Functions that are logically compatible with each other would belong in the same

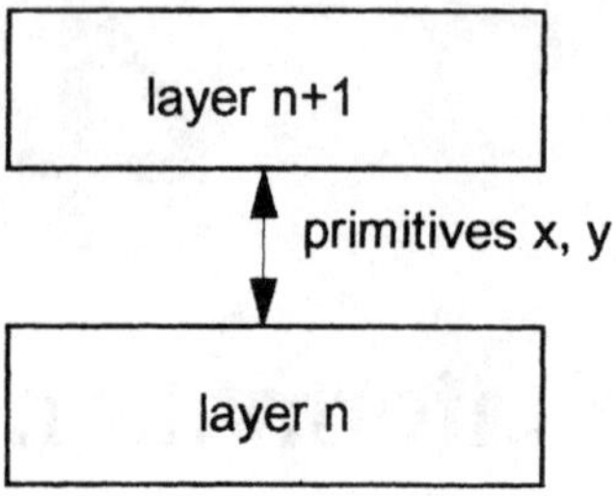

Figure 4.1 Interlayer communication.

set. Each functional set is now called a *layer*. And this leads us to another principle: *layer independence*. Each layer is independent of any other layer in a model. A layer performs designated functions and communicates with other layers through primitives that carry necessary information between layers (see Fig. 4.1).

Systems communicate with each other on a *peer-to-peer basis*. This means that specific layers of one system communicate with their peers in another system. Using these general principles of layering and communication, the International Standards Organization (ISO) developed a specific model for communication with a number of layers performing specific functions. The ISO OSI reference model is shown in Fig. 4.2. ISO created a model consisting of seven layers. As with any other international standard, the reference model was a product of negotiations between a number of involved parties representing different countries and companies. Why seven was chosen instead of six or eight is probably known only to those people who were closely involved, but the resulting functional sets do make sense and are generally accepted in the industry as appropriate.* The functions of each specific layer are discussed in the following sections.

Information is transferred between peer entities and between adjacent layers in various types of protocol data units (PDUs). The application layer generates Application PDUs (APDUs), the presentation layer generates Presentation PDUs (PPDUs), and so on. Peer entities

---

* With some experience gained by various participants in the computer industry, the value of having all seven layers present in the implemented equipment is questioned. Sometimes, it is alleged, the more effective or more efficient implementations are those that omit some of the layers. Usually we find layers 4 through 6, or some portions of them, omitted. There are even attempts in the standards community to create a so-called *skinny stack* of layers. This approach would definitely be cheaper and easier to implement, but what is it going to do to the whole idea of completely open systems? Can the concept survive if some portions of functionality that have been perceived as necessary are not going to be implemented? In such cases, can systems still communicate effectively? These are the questions that the industry is trying to answer, and at the present time there are no agreed answers. So, in this book we will approach the subject of seven layers as something that has the backing of the industry.

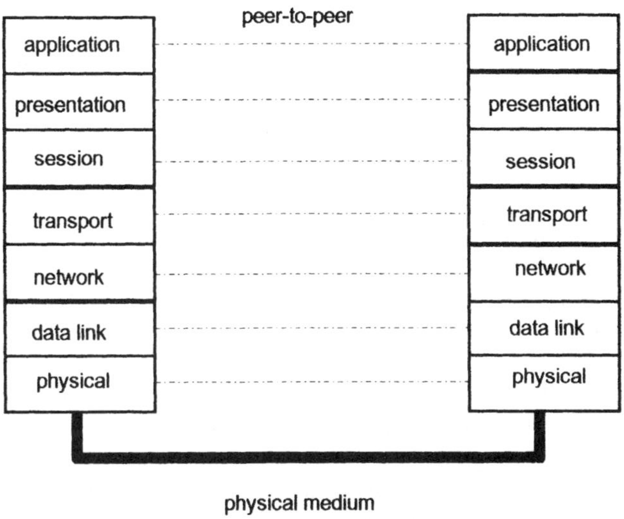

Figure 4.2 ISO OSI reference model.

exchange their own layers' PDUs. Any PDU consists of two parts: user data and protocol control information (PCI). User data is information that must be transferred for the operation of the user's application process. It is supplied either by the user application itself or by an adjacent upper layer (in which case, it is formally called a service data unit—SDU). PCI is control information required for the joint operation of the peer entities in a given layer. It is supplied by each layer. Figure 4.3 shows how a PDU is constructed.

4.1.1 Individual layers

Physical layer (layer 1). The physical layer provides the functions implied in its name: it takes care of all aspects of activation, maintenance, and deactivation of the physical connection. It provides for the transparent bit transfer over the physical connection between involved

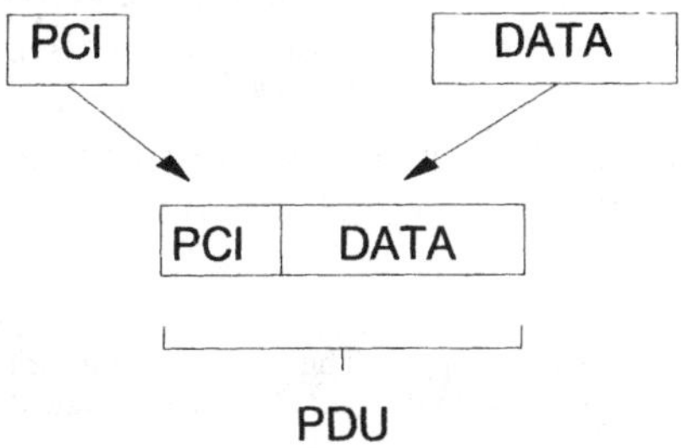

Figure 4.3 PDU construction.

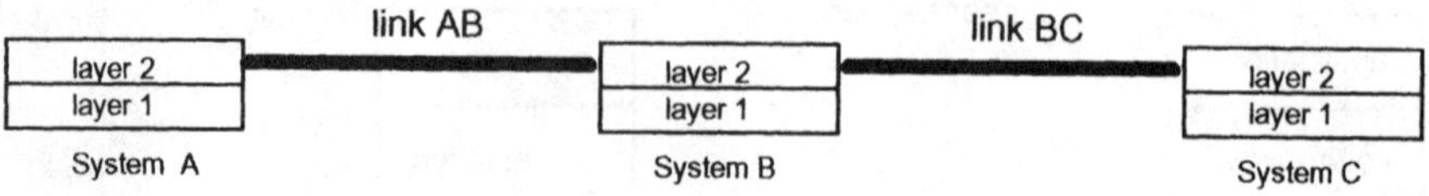

Figure 4.4 Layer 2 scope.

systems. The functions of the physical layer are primarily electrical, which makes this layer unique among its companion layers. This is the only layer that does not massage any data; it just takes care of transmitting and receiving bits of information.*

Data link layer (layer 2). The data link layer provides functions necessary for the establishment, maintenance, and release of data connections over one or several physical connections. It provides a basic framing structure for transferring information between systems. The data link layer is responsible for detecting and correcting transmission errors over a physical link. While layer 1 is transmitting bits of information, the data link layer is overseeing the process, making sure that bits arrive at the destination uncorrupted. If an error is discovered, layer 2 retransmits appropriate data.

Layer 2 operates on a link basis and is not concerned with the delivery of data to the final destination (end system). In Fig. 4.4, the data link layers in open systems A and B are responsible for the link between these two systems only (link AB), while the data link layers in systems B and C are responsible for the link between them (link BC).

Network layer (layer 3). The network layer is responsible for addressing and routing. It provides the functions for the identification of the appropriate routing path through the network. The network layer's operations are independent of the underlying media. Figure 4.5 shows the function of the network layer as a finder of the path through a number of subnetworks, each subnetwork representing an individual network built over a specific communication medium.

Transport layer (layer 4). The transport layer provides functions necessary for the transparent transfer of information between end entities. This layer operates end-to-end, that is, it is concerned with data transfer from the originating entity (system) to the terminating entity. The analogy can be made with layers 1 and 2 in the following way: layer 4

* The physical medium itself is not part of the physical layer. The physical medium provides an actual communication path between systems and is used by the physical layer to transmit data.

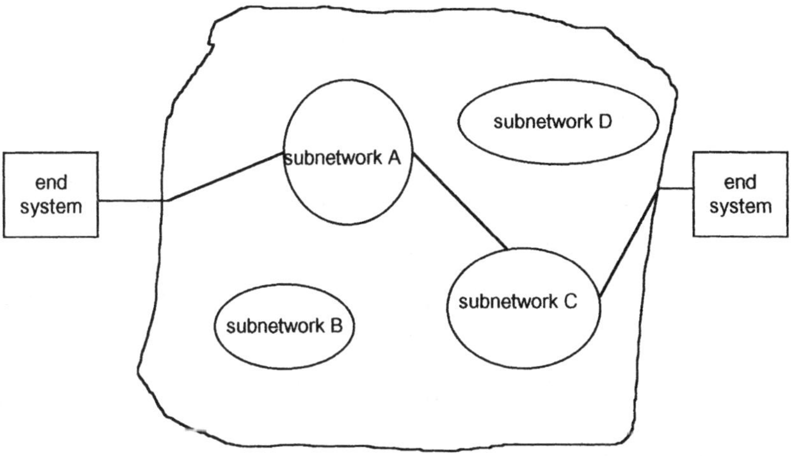

Figure 4.5 Finding a path.

is to layer 3 as layer 2 is to layer 1. While layer 3 is concerned with finding a path through the network, layer 4 ensures that end-to-end information transfer is correct (see Fig. 4.6).

The transport layer is the highest layer in the OSI reference model related to actual transfer of data. The layers above it are applications-oriented and are not involved with the communication medium.

Session layer (layer 5). The session layer provides functions necessary for the synchronization of the dialog between two systems. It establishes, maintains, and releases session connections. It can, for example, keep track of the specific points in the dialog and, if a connection failure occurs, continue a dialog starting with the latest possible point before the breakdown.

Presentation layer (layer 6). The presentation layer provides functions necessary for the negotiation of the representation of information. It is concerned only with representation (syntax) and not with the meaning of information (semantics).

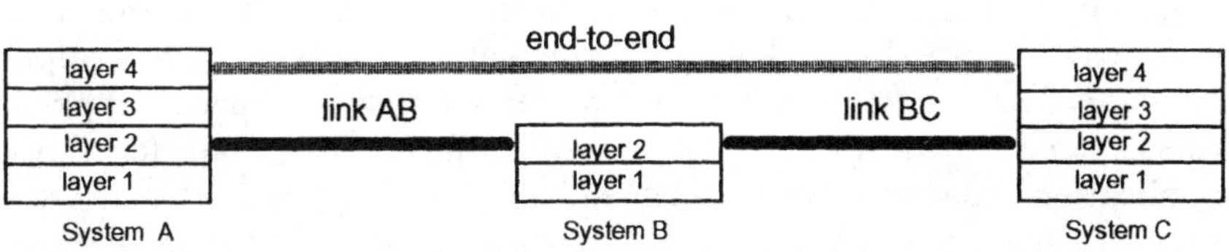

Figure 4.6 Layer 4 scope.

Two main functions are performed by layer 6:

1. Selection of syntax
2. Transformation of syntax (translation)

The functions are similar to those of finding a common language that two people from different countries can understand. For example, a German and a Japanese can't communicate in their native languages because neither one is familiar with the other's native language, but in the process of negotiation they discover that both speak English. So, the negotiated communication language in this case is English.

Application layer (layer 7). The application layer is the highest layer in the OSI model and is specifically concerned with the meaning of information to be transferred. The application layer is the only layer that directly relates to the application process. It performs all communication functions that are not performed by the lower layers. These functions can vary from environment to environment, but some of the common ones are

- Identification of intended communications partners
- Determination of the current availability of communications partners
- Establishment/release of association with applications processes at communications partners
- Establishment of the authority to communicate

and some others of the type mentioned previously.

Since SCAI is specifically an application-layer-related concept, it is useful to describe the structure of the application layer in more detail.

4.1.2 Structure of the Application layer

Layers 1 through 6 have one thing in common: they all perform specific, well-defined functions. This does not mean that all of their intended functions are always used; instead, it means that, if a function is needed, it is used in a specific way. For example, if end-to-end segmentation is required, layer 4 can provide this function.

Layer 7 is different. It is the only layer in the OSI communication stack that is directly involved with application processes (users' applications). Because the application layer must support various applications, it has to be flexible to provide different functions for those applications. Because the number of computer applications will grow with the invention of new services, new functions will have to be introduced in the application layer.

While some common functions within the application layer have been standardized, the majority of currently used or future functions are application-specific. Some of these specific functions may also be used by applications other than those they were created for. This desire and the need to create an environment at the application layer that can be understood and used by many applications developers prompted the standards community to define a structure that would permit a notion of functional building blocks. These building blocks would be created and used for and by the application processes that require them.

The ISO standard specifying the structure of the application layer (ISO 9545) has been evolving since the time when the development of the SCAI standard began. Its evolution is the result of some practical knowledge gained by its designers and implementors. Today's application layer structure standard looks somewhat different from the one used by the SCAI developers. However, the concepts have not changed.*

Since the SCAI standard uses the previously available terminology, the following description of the application layer standard also uses the same terminology. Where appropriate, new terms are identified.

First, the definitions:

Application Process (AP). An AP represents the set of resources, including processing resources, within a real open system that may be used to perform a particular information processing activity.

Application Entity (AE). An AE represents a set of OSI communications capabilities of a particular AP.

Application Service Element (ASE). An ASE is an indivisible component of an AE. It is a combination of application communication functions within an AE that is distinguished for purposes of OSI service and protocol specification.

Single Association Control Function (SACF).† An SACF represents the rules governing the use of ASEs within an SAO. **New term:** *Control Function (CF).*‡

Single Association Object (SAO). A combination of specific ASEs and SACF rules that are used over one particular association. **New term:** *Application Service Object (ASO).*

---

* A reader of the SCAI standard who is familiar with the latest application layer structure standard would find certain discrepancies in the terminology. ISO 9545 contains an *annex* that relates new terminology to the older one.

† A *Multiple Association Control Function (MACF)* is also identified. As its name implies, it is a collection of rules governing the behavior of more than one SAO for the communications needs of a single AP.

‡ A CF concept includes both previous SACF and previous MACF.

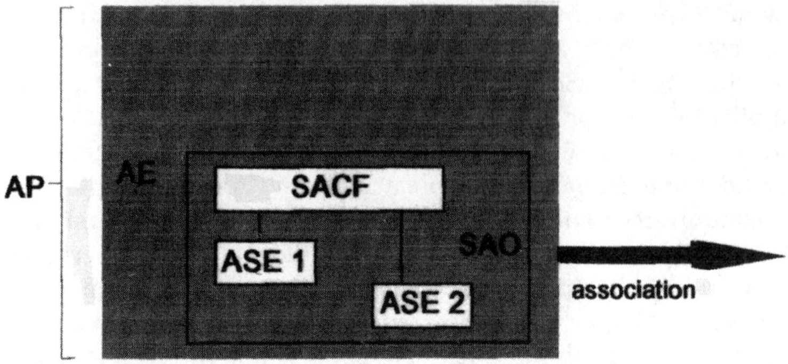

Figure 4.7 Layer 7 structure.

Application Association. A cooperative relationship between two SAO invocations.

Application Context (AC). A set of rules shared in common by two SAO invocations in order to support an application association.

Let's see how it all fits together in Fig. 4.7.

An application process (AP) is a specific application running in a computer and it includes the application layer of the OSI communication stack.* The communication portion of an AP is an application entity (AE). An AE consists of one or more SAOs; each SAO contains a combination of functions collected in one or more ASEs and a set of SACF rules. An association has to be established with a peer AE for the communication to take place. To establish an association, a particular application context (AC), containing a set of rules governing this particular association, is negotiated by the peers.

An AC is a detailed specification of the use of association between two open systems. An AC negotiation allows the selection of a particular set of functions that are applicable for this particular association and available at both sides of the communication path.

4.1.3 How does it all work?

When an application process requires communication with another application process in a remote system, it requests the services of the application layer by triggering the start of a specific procedure (how it is done is not covered in OSI or in the standards associated with it). The application layer establishes an association with its peer in the

* From now on throughout this book, an application process will imply the portion of an AP that is above the AE—the user's process itself, excluding its communications portion.

targeted system. This association is a logical path over which application layer peers will communicate. Establishment of an association requires an exchange of certain information. An established association supports the negotiated application context. We will not discuss this exchange in detail here since the purpose of this section is to outline general common actions that take place within the seven-layer model. It is sufficient to realize that the establishment of an association is a necessary action.

When an association is established, it can be used for peer-to-peer communication. The communication is performed through the use of protocol elements (PDUs). Each PDU carries specific commands, controls, and/or information. For example, a PDU may carry a request to perform a search of a specific name, or it may carry information necessary to complete a specific procedure. The purpose of an application PDU is to relate information to the peer application (*horizontal communication*), but this PDU must first travel down the seven-layer stack (*vertical communication*) to reach a physical medium over which an instance of communication will take place.

The created Application PDU is transferred down to the presentation layer. Since the presentation layer communicates with its peer, the presentation layer control information is added to the application PDU and it becomes a presentation PDU. This PDU is transferred further down to layer 5.

The process of specific-layer PDU creation takes place at each layer, and, finally, layer 1 transmits data over the physical medium to the destination system. So, what is being transmitted? The information that is useful for the application process is created and analyzed at the

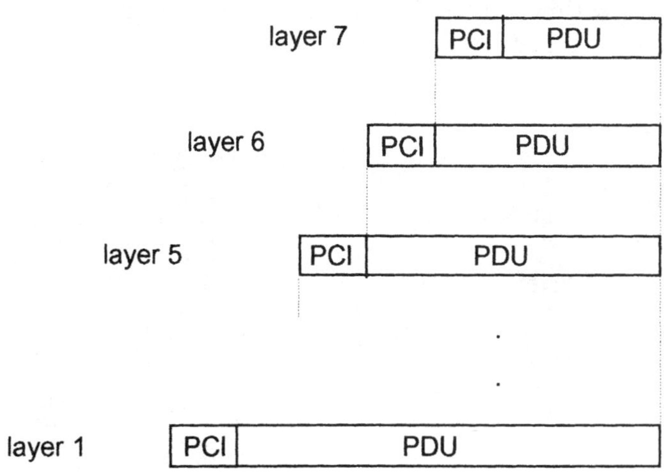

Figure 4.8 PDU transformation.

application layer, but each layer adds its specific information which is necessary for the successful communication with a remote system. Figure 4.8 schematically shows how the information content grows as the PDU passes through the layers down to the physical medium for transmission. The process that takes place at the remote (receiving) system is a reverse of the one just described. The received information package is stripped of the appropriate control information at each layer until the user data at layer 7 is exactly as it was originally created at the originating system. Now this information is translated into a specific parameter (or parameters) that is passed up to the application process.

4.1.4 OSI primitives

There is one more aspect of OSI that is worth mentioning because it is used in the SCAI standard to describe functions. While each PDU is targeted for the peer in the same layer, it is transported to the peer system over the physical media (under the control of the physical layer). The same PDU is transferred up or down the OSI stack in the form of one of four service primitives defined for this purpose:

1. *request* (req)
2. *indication* (ind)
3. *response* (rsp)
4. *confirm* (cnf)

Figure 4.9 shows the usage of these primitives.

System A originates a primitive *request* which is passed to the destination system B. Within system B, this information is considered a primitive *indication*. It is delivered to the peer of the entity that origi-

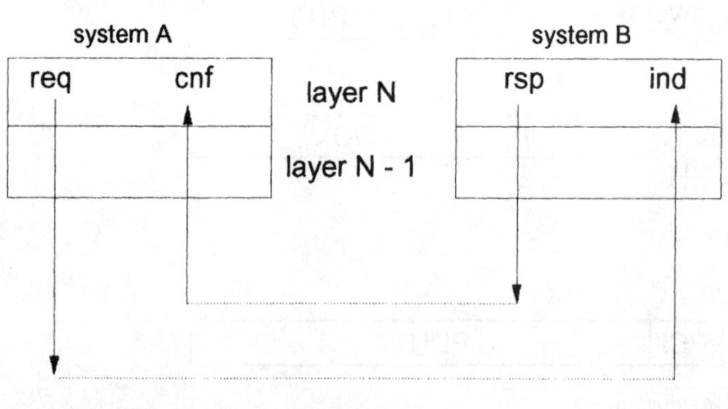

Figure 4.9 OSI service primitives.

nated this exchange. After the peer entity analyzes information in the primitive, it issues a *response* primitive containing response information. When it is passed to system A, it is considered a *confirm* primitive and is delivered to the originating peer entity.

A service can be confirmed or unconfirmed. A confirmed service requires all four primitives. The result of the activation of a confirmed primitive is a handshake between the two peer systems. An unconfirmed service requires only two primitives: *request* and *indication.*

Usually, information (parameters) contained in a pair *req/ind* is the same. It is also true for a pair *rsp/cnf.* However, there can be instances when it is not true. The model, therefore, is general and permits variation of parameters within the pair of primitives. Consider a fictitious service *Do.* In one case, all parameters of the *Do-indication* primitive are the same as of the *Do-request* primitive (Table 4.1).

In another case, there is a need for an extra parameter in the *Do-indication* primitive (Table 4.2).

The additional parameter *how* is used in the *Do-indication* primitive to specify more detail about the service. It may be that each system that can perform the *Do* service does it in a different way, and an additional parameter is generated by this system itself. The conventions for indicating the presence of a parameter are as follows:

M is mandatory—the parameter must be present.

U is optional (by a user)—the parameter may be present.

C is conditional—the parameter is present in *ind* or *cnf* if it is present in *req* or *rsp.*

(=) is equality of the *ind* or *cnf* to *req* or *rsp.*

The naming convention for primitives is:

```
layer-name, service, service primitive
```

TABLE 4.1

parameter	request	indication
what	M	M(=)
when	M	M(=)

TABLE 4.2

parameter	request	indication
what	M	M(=)
when	M	M(=)
how		M

For example, if a *Do* service is invoked at layer 7, the associated PDU is named:

```
A-Do.request
```

where A stands for application layer.

To relate this concept to the notion of PDUs discussed previously, just imagine that when a service is invoked at the application layer, an APDU is generated and is passed through the layers to the peer entity as *request* and then *indication* primitives.

4.2 ASN.1

Abstract Syntax Notation One (ASN.1) is a language that is used to represent information in the abstract form. ASN.1 can be compared with a high-level programming language, such as C or Pascal. It allows the description of information without considering the details of the implementation that will use this information. ASN.1 is used in the definition of communications protocols. Although a protocol at any layer can be described with ASN.1, its primary use is for the application and presentation layer protocols.

Two concepts are important to understand in relation to ASN.1:

- *abstract syntax*
- *transfer syntax*

An abstract syntax defines data types and values for these types. For example, if we need to communicate information about employees, we could define an employee record as follows:

```
EMPLOYEE:=
        Name
        Age
        Gender
        Address
```

The employee record constitutes, in ASN.1 terms, a data type. All we have defined so far is the type of information that we need to communicate. Of course, nobody needs to communicate empty records; what can be expected is a record of a specific individual. When we fill the fields of the record with meaningful data, the value of the data type is assigned.

```
EMPLOYEE::=
        Name = Smith
        Age = 40
        Gender = male
        Address = somewhere
```

We now have an abstract syntax for employee records. The abstract syntax does not specify how the actual information message is coded or how it is to be implemented. It does define the data type for this record. Referencing this type would make its components visible. The values of this type are also defined. One of the possible values is given in the example, but an indefinite number of other values is possible. The value of this data type consists of four elements. By giving each element (e.g., *Address*) a description (e.g., element *Name* now has a description *Smith*), we are assigning its value (*Smith* is a value of *Name* in this case). These descriptions can be of two kinds: (1) when a specific set of values is permitted (e.g., *Gender* can be either male or female) and (2) when a range is defined (e.g., *Name* is constructed out of letters only).

A transfer syntax defines an actual representation of the values assigned by the abstract syntax. In our example of the abstract syntax, we indicated a possible value, but did not mention how this particular value is to be represented in an actual message. It is the job of the presentation layer to negotiate the transfer syntax. The presentation layer, through the exchange of messages between peer systems, is supposed to identify a specific transfer syntax out of many theoretically possible syntaxes. In reality, though, there is only one standardized transfer syntax that is currently used. This syntax is created by applying Basic Encoding Rules (BER) to an abstract syntax. When BER is applied to the data type values of the abstract syntax, a specific encoding of these values is achieved. Information thus encoded is transferred between peer systems and it is unambiguous because both sides agreed on the method for encoding.

4.2.1 ASN.1 fundamentals

ASN.1, like any other high-level language, allows creation of modules containing necessary definitions. These modules are created arbitrarily and can cross-reference each other. The syntax of a module is as follows:

```
Module-name {Module-ID} DEFINITIONS::=
BEGIN
        Imports/Exports
        Declarations
END
```

Each module begins with its name. The name is followed by the value that identifies this module. We will discuss these values later. The body of the module is contained between BEGIN and END. In the body, the data types and values that can be imported from other modules are listed first. Then, data types and values that can be exported to other modules (used by other modules) are listed. The main part of the module's body contains declarations of the data types and values defined by this module.

If we wanted to define the Employee data type from our previous example as a module, the following would be the result:

```
Employee {OBJECT IDENTIFIER} DEFINITIONS::=
BEGIN
{
  name::=OCTET STRING,
  age::=INTEGER,
  gender::=BOOLEAN,
  address::=OCTET STRING}
END
```

ASN.1 allows two kinds of data type—simple and structured. Both kinds are used in the example and we will discuss them momentarily. The following are the simple data types:

INTEGER—represents whole numbers (1, 2, 1890, –38, etc.).

REAL—represents rational numbers (corresponds to the floating-point type of programming language).

NULL—represents *null.*

BOOLEAN—represents logical true and false.

ENUMERATED—represents a list of assigned integers. For example:

```
Colors::=ENUMERATED
{
        blue(1), red(2), green(3), white(4)}
```

BIT STRING—represents a sequence of bits.

OCTET STRING—represents a sequence of octets.

OBJECT IDENTIFIER—represents the value that uniquely identifies an ASN.1 object (module, etc.).

While all other simple types are easy enough to understand with a little knowledge of programming languages, the object identifier type is different. It is specific to ASN.1 because it serves as a method of identifying various ASN.1 objects, for example, modules. Each ASN.1 object requiring a unique reference is assigned an *object identifier* (*ID*) by an appropriate standards body. The IDs are assigned according to the node in the *object ID tree* where the object belongs. An essential part of the object ID tree is shown in Fig. 4.10.

The object ID tree has three main branches: ITU-T, ISO, and Joint-ISO-ITU-T. The value after each node name is the object ID value for that node. The ITU-T branch consists of four alternatives:

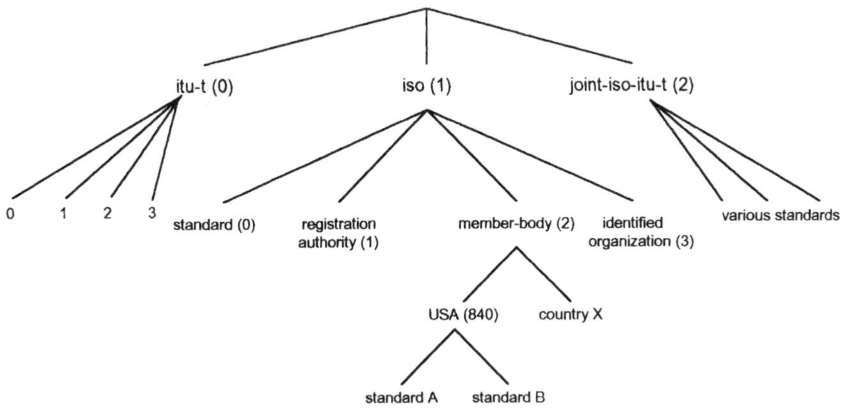

Figure 4.10 Object ID tree.

- value 0—for ITU-T recommendations (e.g., Q.931)
- value 1—for ITU-T questions (e.g., new switching techniques)*
- value 2—for administration (e.g., Germany)
- value 3—for network operators (e.g., Telecom Canada)

The Joint-ITUT-ISO branch consists of many alternatives corresponding to various standards (e.g., ACSE).

The ISO branch has four alternatives:

- *Standard*—for ISO standards (e.g., 9545—Application Layer Structure)
- *Registration authority*—for specific OSI registration authorities
- *Member body*—for country members of ISO (e.g., U.S.A.)
- *Identified organizations*—for the identification of any organization

An object ID value consists of the values of all nodes that have to be traversed; for example, the object ID of the fictional U.S. standard B is equal to

```
{iso(1) member-body(2) uSA(840) standardB (ZZZ)}
```

The structured types are those that are defined in terms of simple types. Structured types allow protocol designers to create types corresponding to specific needs. These complex (structured) types are then referenced in the same manner as a simple type. For example, to define

* An ITU-T question is a work item that ITU-T has agreed to work on. Each question has an identification number.

a structured type HOUSE that consists of many components (doors, closets, roof, and so on), we can reference this type as a collection of elements, rather than referencing the elements themselves. The following are the ASN.1 structured types:

SET—represents a collection of component types; each component has its own type, simple or structured. The HOUSE type example can serve as an example of the SET type.

```
House::=SET
{
  door    ENUMERATED,
  closets INTEGER,
  ......
  roof    BOOLEAN}
```

Now, when we need a data type containing all the preceding components, we can simply use HOUSE for this purpose.

SEQUENCE—represents a collection of component types, just like SET; the difference is that the SET type does not imply any ordering of components, while the SEQUENCE type defines the ordered sequence of components. In our example, if the HOUSE type is defined in terms of SEQUENCE, it would mandate the sequence of components.

SET OF—represents a single component type with an unordered list of values associated with that type. For example:

```
Number ::= SET OF INTEGER
```

SEQUENCE OF—represents a single component type with an ordered list of values. Again, the difference from the SET OF type is a characteristic of orderliness of the members of the set and sequence.

CHOICE—represents a collection of components that are alternatives of distinct types. For example:

```
Number ::= CHOICE
{
      a      INTEGER,
      b      REAL,
      c      NULL}
```

Only one of the three components of the type number can represent its value in a given situation.

ANY—represents any defined data type and, therefore, presents a problem for the developers, unless it is accompanied by the additional construct: DEFINED BY. In most cases, the type ANY is used with DEFINED BY, where the latter part points to a specific definition of the type.

ASN.1 employs the concept of tagging. Tags are values assigned to specific types to differentiate among them. There are four general classes of tags:

1. *Universal tag.* A universal tag is assigned to all types that must be globally unique. Types INTEGER, SEQUENCE, NULL have universal tags assigned to them—respectively 2, 16, and 5. All ASN.1 built-in data types have universal tags assigned to them.
2. *Application-wide tag.* This kind of tag is intended for use within an application. As envisioned, data types for a specific application would be tagged with application-wide tags and, therefore, these data types would be unique within this application only.
3. *Context-specific tag.* This kind of tag is used to uniquely identify components of a structured data type within this data type. For example:

   ```
   Car ::= CHOICE
   {
           ford   [0] INTEGER,
           nissan [1] INTEGER,
           fiat   [2] INTEGER}
   ```

   Since all components of the data type CAR are of the same type, the tags are used to differentiate among them.
4. *Private tag.* This kind of tag is envisioned for private companies that desire to extend a standard and create their own variation of it. Private tags are unique within the environment of the private company. By using private tags, the company in question is essentially deploying a nonstandard solution in the purest sense of the word.

Before we consider an example of an ASN.1 module with the elements presented so far, there is one more item that has to be mentioned. When data types that correspond to elements of communication protocols are created, there is a need to indicate that certain parameters in the protocol elements are not always present. This requirement is fulfilled in ASN.1 through the use of the keyword OPTIONAL. Placed after a data type component, it indicates that this component may not be present in a given situation.

Now, let's consider the simple ASN.1 example shown in Fig. 4.11.

In the example, a module "Personnel" with an object ID of (2 9 5 6 1) is defined to communicate personnel records and available jobs. The module defines two data types: "Employee" and "Jobs." The data type Employee is defined as a set of instances of another data type—Record. This data type contains four components, two of which—gender and age—are defined using built-in types, BOOLEAN and INTEGER. The other component—address—is defined by data type Whereisit, which

```
Personnel {2, 9, 5, 6, 1} DEFINITIONS::=
BEGIN
IMPORTS
  Whereisit
  FROM Other-Module {2, 9, 5, 6, 5};
EXPORTS
Employname;
        Employee ::= SET OF Record
        Record ::= SEQUENCE
        {
               name        Employname,
               age         INTEGER,
               gender      BOOLEAN,
               address          Whereisit}
        Employname ::= SEQUENCE
        {
               firstname  [0]  OCTET STRING   OPTIONAL,
               lastname   [1]  OCTET STRING}
        Jobs ::= ENUMERATED
        {
               jobA        (0),
               jobB        (1),
               jobC        (2),
               jobD        (3)}
END
```

Figure 4.11 ASN.1 example.

is imported from another module, Other-Module, identified by the object ID with value that equals (2 9 5 6 5). This means that Whereisit is defined in Other-Module. The last component in the type Record—name—is of the type Employname, which is defined further in the same module as a sequence of a first name and a last name. Both of these components are of the type OCTET STRING, and the first name is an optional component. Since both components are of the same type, they are tagged with context-specific tags to uniquely identify each of them within this particular data structure. In addition, the data type Employname is exported, which means it is visible to other ASN.1 modules and can be referenced by them.

The data structure Jobs is of Enumerated type; it lists four jobs with integer values assigned to them for the differentiation purposes.

4.2.2 BER

Basic Encoding Rules (BER) are used to transform an abstract syntax defined by an application into a transfer syntax which will actually be transferred to the peer system. Although there can be many encoding rules for this purpose, BER is an international standard and the only one created for this purpose so far. Since it is possible to have different transfer syntaxes derived from the same abstract syntax, the presentation layer performs negotiation of the transfer syntax for an application association.

BER is a set of rules of encoding of data type values. A simple example of different representations of the same data is a number represented in a decimal or a binary form. Number "ten" is represented as "10" (decimal) or "1010" (binary). The information is the same, its representation is different. BER contains rules for encoding each data type value that ASN.1 uses. The general approach to encoding employs the structure *identifier-length-content* (see Fig. 4.12).

Each value is encoded using this structure. The identifier part conveys the following information:

- the tag class (universal, application-wide, etc.)
- the tag value (integer, real, etc.)
- the form of the encoding

The form of the encoding can be either primitive or constructed. Primitive encoding of the content means that the value of the content is encoded in a series of octets. If the form of the encoding is constructed, the content is a series of nested encodings as in Fig. 4.13.

The identifier occupies one or more octets. A single octet identifier is shown in Fig. 4.14.

Figure 4.12 Basic encoding of information.

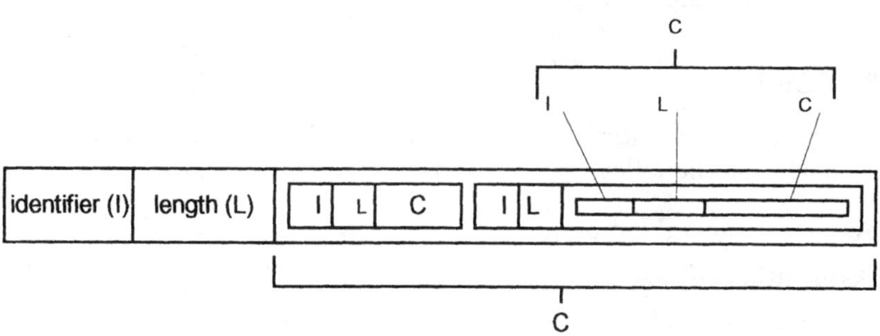

Figure 4.13 Constructed form.

8	7	6	5	4	3	2	1	bits
A	A	F	V	V	V	V	V	

Figure 4.14 Identifier.

Bits 7 and 8 (represented by A) specify the class of the tag; bit 6 (F) specifies the form of the encoding; bits 1 through 5 (Vs) specify the value of the tag. If all bits 1 through 5 are ones, they indicate that the identifier is more than one octet in length and the following octets specify the value of the tag. The identifier value "00000010" represents a universal integer primitively encoded. The identifier value "10110000" represents a context-specific constructed sequence.

The length identifies the length of the content. It can be either short, long, or indefinite. If the length is indefinite, the end of the content is indicated by two octets of zeros. The indefinite length is used only with the constructed form of content. A short length is used to indicate that the content size is no more than 127 octets. For a greater size a long form of length is used.

The short form of length occupies a single octet. Bit 8 is always a zero (this is an indication of the short length) and the other seven bits represent the value, which is the number of octets. For example, the length value "00011101" identifies that there are 29 octets of content following the length.

The long form of length occupies more than one octet. Bit 8 in the first octet is always a 1 (this is an indication of the long length). The other seven bits of the first octet indicate the size of the length. The value of the first octet, "10010100," indicates that there are 20 octets following the first one that represent the actual length of the content (must be a very long data stream!).

The content part of the encoding is coded according to the rules for each type—integer, real, sequence, and so forth. I do not intend to discuss the encoding of each type here; it is sufficient to understand the encoding principles and general rules.

4.3 ACSE

The Association Control Service Element (ACSE) is a standard ASE that performs the following functions:

- Association establishment
- Association release
- Association abort
- Authentication

Association establishment and release functions are invoked in the very beginning and at the very end of the communication activity between two open systems. The abort service is invoked when a user wants to discontinue the communications activity or when the underlying protocol stack discovers an unrecoverable error.

Authentication is a security function added to the ACSE standard relatively recently. Whether this function is used or not depends on the requirements of the particular application environment.

The ACSE is used by all open systems because all communications must be preceded by an association establishment (in other words, a "let's talk" agreement must be made before a real conversation begins). Association establishment is an important element of the open system communication because a great deal of identification information is exchanged during an association establishment. It is, therefore, necessary that this exchange be performed in a standard manner to be recognized and understood by all open systems.

Examples of exchanged identification information are the names of application processes invoking the action and the name of an application context. These names are referred to as object identifiers.

4.4 ROSE

The Remote Operations Service Element (ROSE) is another standardized ASE that is widely used in the distributed applications environment. The ROSE's purpose in life is to provide for the remote invocation of operations. When a system needs to request a service from another (remote) system, it sends the remote system a request for that service. The remote system accepts or does not accept the request and responds accordingly. ROSE provides a standard mechanism for exchanging necessary information between systems to accomplish distributed processing. Let's look at the ROSE operations model in Fig. 4.15.

The remote operations model is simple: a request to perform an operation is replied to with an indication of a success or a failure. There are five classes of operation in ROSE, as shown in Table 4.3.

Operations invoked by one AE (the invoker) are performed by the other AE (the performer).

Figure 4.15 Remote operations model.

TABLE 4.3

Operation Class 1	Synchronous, reporting success or failure (result or error)
Operation Class 2	Asynchronous, reporting success or failure
Operation Class 3	Asynchronous, reporting failure only, if any
Operation Class 4	Asynchronous, reporting success only
Operation Class 5	Asynchronous, outcome not reported

The following ROSE services are available:

- RO-INVOKE
- RO-RESULT
- RO-ERROR
- RO-REJECT-U
- RO-REJECT-P

The notation RO is used for all ROSE services.

An invoking AE uses RO-INVOKE to send a request to perform a certain operation by a performing AE.

A performing AE uses RO-RESULT to report the success of a requested operation to an invoking AE.

The RO-ERROR is used by the performing AE to report an unsuccessful operation to an invoking AE.

Both RO-REJECTs are used in the cases when the request or reply cannot be accepted due to discovered problems. The RO-REJECT-U is used when the ROSE user has detected a problem, and the RO-REJECT-P is used when the ROSE provider has detected a problem.

A ROSE user is a user of ROSE services, while a ROSE provider is a provider of ROSE services.

When a ROSE service is requested, a specific ROSE APDU is generated. For example, when an operation is requesting an invoke service, an invoke APDU (ROIV) is generated.

4.5 Object-oriented Methodology

Object-oriented methodology to modeling and design is a way of approaching a problem using abstract models. The more traditional way of thinking about a problem used to be functional. In the functional approach, an analysis of the desired functions is performed and then a product (a communication protocol, a computer program, or the like) is designed based on the understanding of the functions. In the object-oriented approach, the real-life objects perceived to be a part of the problem space are first analyzed and described. The solution is based on knowledge of these objects.

Object-oriented methodology is described by various users of this technique in different ways. They use different notations and terminology, but the concepts are the same. Lack of standard notation is a problem similar to that of two people speaking similar, but not the same, languages and trying to gauge whether the word means the same thing in both languages or not. Nevertheless, this is the state of affairs in the object-oriented world, and the industry is attempting to make the best of it.

I do not intend to describe object-oriented methodology and techniques in a comprehensive way. Instead, I want to provide the reader with the general background of this methodology. The terminology used here is the same as was applied for the description of the objects in the SCAI problem space.

An object-oriented approach implies that the problem space is organized as a collection of objects with their specific behavior and associated data. Understanding these objects in a simplified form promotes a better understanding of the entire problem and, therefore, a cleaner solution for the problem.

An *object* is a concrete distinguishable entity. Examples of objects are a tree, a telephone, a car. Objects are classified according to their characteristics. The objects that have the same characteristics belong to the same *class*. Each object is an instance of its class. The selection of a class depends on the characteristics that a particular application is interested in. For example, we can define a class "animals," with each known animal being an instance of this class. This would imply that a common characteristic of interest to an application is the fact that animals are alive. If, on the other hand, an application is interested in animals with four legs only, then we would define a class "four-legged animals" and would not care about the existence of other animals.

A class can be subdivided into *subclasses,* which constitute a further refinement of a class into more specific groups. In our previous exam-

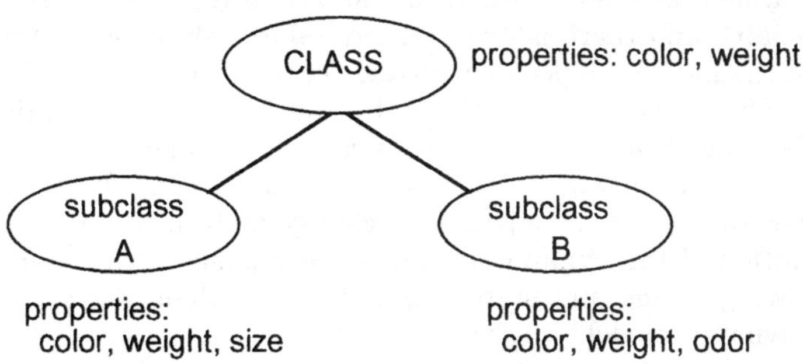

Figure 4.16 Inheritance.

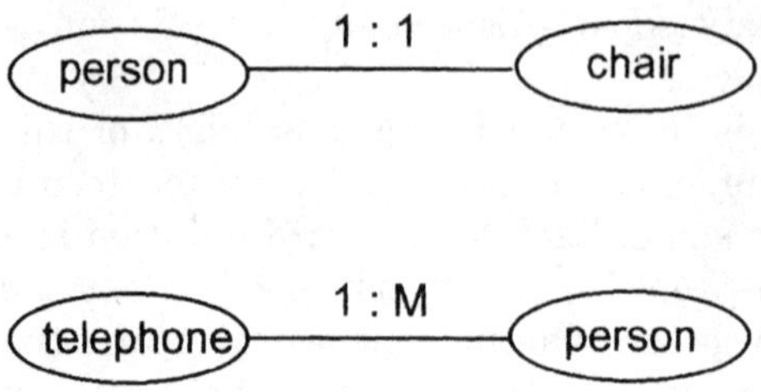

Figure 4.17 Relationships of objects.

ple of the class "animals," we can imagine subclasses "mammals," "birds," and so on, if this classification is useful to the application, or subclasses "small animals," "medium animals," and "large animals," if this is what the application is interested in.

Tied in with the concept of subclasses is the concept of *inheritance.* Inheritance is the sharing of common properties by class members based on a hierarchical relationship. All properties of a class are inherited by its subclasses. A subclass also has additional properties; this is how one subclass is distinguished from another subclass of the same class.

Figure 4.16 shows that two subclasses inherit two properties of the class (color and weight), and each has an additional property: subclass A has the property of size and subclass B has the property of odor.

Objects are described with attributes and operations. *Attributes* are specific characteristics of a class of objects (e.g., color, age, profession). *Operations* are actions associated with an object class. The behavior of the object is another way of expressing the same property of actions. For example, the operations of a television set are turn on, turn off, switch channel, and so on. In the object-oriented methodology, operations are *polymorphic,* that is, they can behave differently for different classes. Operation "turn on" behaves differently for a TV set and an automobile. A specific implementation of an operation for a specific class is called a *method.* Now we can say that an object is characterized by its attributes and methods.

Classes of objects can relate to each other; for example, there is a one-to-one relationship between a person and a chair (only one person can sit in a chair), and there is one-to-many relationship between a telephone and the number of people who can use it (see Fig. 4.17).

We can identify two types of models: static and dynamic. The static model describes the object, its characteristics, and its relationship to other objects. This is usually called an *object model.* The dynamic model describes the aspects of objects that change with time. Usually, the representation of the dynamic model is accomplished with state diagrams showing states and state transitions. A call model is an example of a dynamic model.

Chapter

5

The SCAI Standard

In the earlier chapters we identified SCAI as a technological solution to the issue of computer/telecom integration. This chapter deals with SCAI as a standard.

The SCAI standard, T1.626, has been developed by the ANSI Committee T1. T1 (T stands for telecommunications) is chartered with the development of the U.S. telecommunications standards and is involved with the development of ISDN, Intelligent Networks (IN), Personal Communications Services (PCS), and other standards. For a more detailed description of the responsibilities, rules, and procedures of T1, see Chap. 12. The SCAI standard has been developed by the ISDN Networking subworking group of T1S1.1 and approved by T1 in November 1992. To order the SCAI standard, either call ANSI at 212-642-4900 or write:

The American National Standards Institute Inc.
11 West 42nd Street, 13th Floor
New York, NY 10036

Sections 5.2 through 5.8 address different aspects of the standard T1.626, SCAI. Since work on SCAI continues in T1S1 and international standards committees, enhancements and additions are expected to the existing standard, but these are only potential changes. Still, I would like to mention the areas of possible change to alert the reader to what can be expected. To do this, every chapter that contains items that may be changed addresses the possibility of a change under the heading "**In the works:**".

5.1 Why Is a Standard Needed?

This is a good question and people involved with the development of standards are asking this question, too. Everybody is familiar with standards in daily life: electrical plugs, light bulbs, beds—these are examples of standards that were created for the benefit of consumers and for the ease of manufacturing and installation. When we buy an electrical appliance in any American store, we know that we can plug it in to any electrical outlet at home (but not in Europe, because standards in Europe are different!). The same idea is valid for more complex systems used in computing and communications. Wouldn't it be nice to buy any computer and run any application software on it? Wouldn't it be nice to be able to communicate with your friend's remote computer without asking the friend what kind of computer it is and whether certain parameters are set in a certain way? In order to achieve these goals, standards are created.

There are two ways in which standards are developed:

1. Through some kind of cooperation in the industry, e.g., ISDN
2. Through a de facto standardization process—for example, Windows—which is, depending on who is talking, an operating system, a user interface, or whatever.

Although eventually the market may decide which approaches will become de facto standards, it may take a long time, or it may never happen. An industry—in this case, a computer/telecommunication industry—can decide what kind of standards should be developed. The industry makes these kinds of decisions based on its perception of the marketplace, that is, what is good for the users of the technology. The users, though, participate in the process only indirectly through market research conducted by manufacturers, service providers, and others. But, since it is the user who ultimately pays (or does not pay!) for any technology and services, standardization decisions are usually based on the industry's perception of this user.

So, why is standardization good for the users and the industry? First, a standard provides users with the possibility of a multivendor environment. No need to rely on one vendor (manufacturer) if the vendor does not perform up to the user's expectations. If an implementation of a system/network is accomplished in a standard way, another vendor's equipment can be added or substituted without a great deal of pain.

Second, for complex concepts and solutions, standards provide a common foundation, based on which the evolution of a specific concept or technology may take place, making the lives of everybody involved a bit easier.

Third, manufacturers with standard implementations have the upper hand in selling their equipment. In addition, a software manufacturer can develop software once and then port it to any machine. This saves a lot of resources.

And fourth, deliberations during standards creation stimulate thoughts in the industry and, at the same time, raise interesting technical issues that otherwise might remain hidden from the public.

There are serious drawbacks in the standardization process. To start, it is a slow process. Assurance of fairness is often the key reason for slow development. Any participating organization can object to one or another aspect of the developing standard and it takes a long time to either resolve an objection or to make sure that the vast majority of the participants disagree with the objection while an objecting organization exhausts its efforts to make a change. When a standard is ready for approval there are several stages to go through, again to assure fairness.

Another problem with standards is that, typically, no standard is completely deterministic. Because creation of standards involves compromise, agreements are often made to allow two options for the same function to be standardized (sometimes, more than two). Existing equipment also influences decisions by requiring that certain capabilities/functions be performed in a certain way to allow older equipment to be covered by new standards.

Some issues are not discussed in the standards committees because they have direct business or legal implications. For example, how services are to be billed is not discussed in the standards committees. Instead, ITU-T (formerly CCITT) discusses *charging principles*—common capabilities to support charging. Direct involvement in billing issues is discouraged. And, finally, companies involved with the development of standards do have strategic interests to move a standard one way or another. Because the standards arena is where many competitors meet, their discussions are not only time-consuming, but, at times, quite heated.

Differing interests clash during the process of standards creation and the process is not perfect, but this is an open and fair way of creating something that will benefit our society: standards for communication and computing. Until a new process is invented, the one that was used to create standards like SCAI is here to stay.

5.2 SCAI Principles

Very early in the development of the SCAI standard it was recognized that SCAI could cover a wide spectrum of possibilities and its development would require the identification of principles that would drive such development. These principles were identified based on the intended use of SCAI and technological directions in the industry.

Figure 5.1 Transport for SCAI information.

5.2.1 OSI compatibility

Development of a protocol architecture and a protocol requires some basis. It is necessary to decide how particular work relates to existing and developing interfaces and protocols. Since many developing protocols and concepts are based on OSI (for example, network management, intelligent networks, directory services), it was decided that OSI would be the basis for the development of the SCAI standard. The principle of compatibility with OSI has several aspects.

Transport independence. Since SCAI is an application layer protocol (see Secs. 4.1 and 5.3.3), it is independent of the underlying layers (1 through 6). This allows any of the existing or newly developed suites of protocols to be used as transport mechanisms for SCAI. A transport mechanism implies a protocol that can be used to transport SCAI information between communicating systems (see Fig. 5.1). SCAI information is any operational data exchanged for the purposes of a SCAI application. Existing protocols like X.25, DSS1, SS7,* and other current and future protocols capable of carrying application layer information can be used as transport mechanisms for SCAI. Therefore, SCAI standard development is concentrated only on application-level functionality. Of course, it is one thing to define an independent standard like SCAI and another thing to deploy it in the real customer's environment. The implementor has to deal with the issue of selection of the appropriate protocol suite (for example, SCAI or X.25). The SCAI standard, however, allows an implementor not to be tied to a particular transport mechanism at the time of initial deployment or in the future, when new transport mechanisms appear.

Layer 7 structure. The development of the SCAI protocol follows the structure of the application layer. This allows merging of multiple applications/services on the same system if the system is implemented based on OSI concepts.

---

* See Chap. 7 for more on these protocols.

Reusability of functions. The intent is to create and identify functions and functional sets in a way that allows their reusability by the same applications or by other applications through their inclusion in a specific *application context*. If an ASE has been created for a purpose, such as for SCAI, it can be used outside of SCAI by being included in the application context identified for a different purpose (different application).

Utilization of existing standardized functions where appropriate. This principle serves two purposes: (1) it allows compatibility with the rest of the OSI world as it relates to already standardized functions, and (2) it saves time because it inhibits a tendency to "recreate (a better!) wheel." Two existing standard functional sets were identified as necessary for the SCAI standard: ACSE and ROSE.

5.2.2 Support for private and public networks

ITU-T (the world's leading standards organization in the area of telecommunications) defines public networks as those serving the general public with no restrictions, and defines private networks as those serving specific sets of customers. Without going into a dissertation on the merits of this definition, we can clearly identify public networks in the United States as local telephone companies (such as New Jersey Bell) and long distance companies (such as AT&T). Examples of private networks would be corporate PBX networks. The question for the SCAI standard developers was whether there are differences significant enough between the two types of networks to warrant different standards for each of them. The answer to this question is that the requirements for both public and private networks are either the same or very similar, in most cases. The principle, therefore, was identified to develop a standard supporting both types of networks.

5.2.3 Close relationship with the ISDN services*

Since SCAI operates in the telecommunications environment, it has to support telecommunications services. Although it is impossible to identify all telecommunications services that SCAI should support, some of them are obvious, especially those offered by telecommunica-

* The phrase "close relationship" is obviously vague. This was done because it was impossible to precisely identify the relationship of ISDN and SCAI. It was clear, though, that this relationship is important for compatibility and evolution reasons and that it is necessary to base certain ideas on ISDN. Given the environment of the standards work with its goal of consensus, this phrase was the best that could be agreed on, but the fact of the existence of such a close relationship was clearly stated and later used in the standards development work.

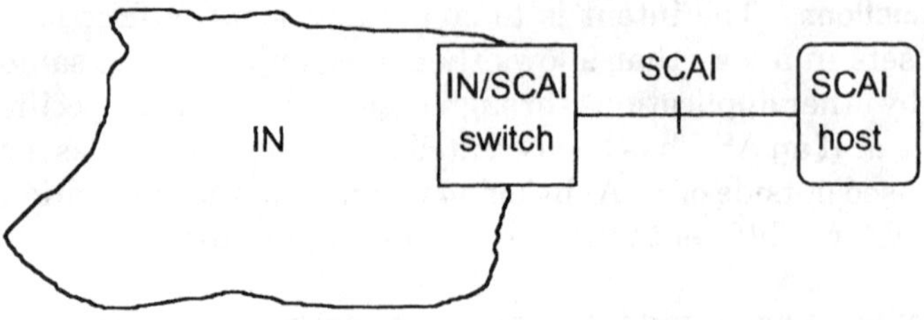

Figure 5.2 IN/SCAI relationship.

tions networks today. The question is how these telecommunication services are to be defined. A principle was identified that the development of a telecommunications service for SCAI will begin with the ISDN service description, if one exists. If the ISDN service description completely corresponds to the requirements of the SCAI service, it is to be used in the SCAI standard. If needed, this description is modified by additional SCAI requirements. In case no ISDN service has been defined, a new SCAI description will be created. For example, there exists an ISDN supplementary service, *Call Transfer*. This ISDN service is identical to the one envisioned for SCAI. Therefore, its description was adopted.

5.2.4 Close relationship with IN*

Intelligent Network (IN) is a new network concept that is being developed and standardized. Its main premise is distribution of intelligence in the networks away from switches to other network entities—host computers, databases, and so on. The IN concept identifies functions of specific network entities and interfaces between them. A switching system (*switch* in SCAI terminology) is one of those entities, and it is the one responsible for call processing (although in the Intelligent Network, a switch communicates with other network entities to make some of the call processing decisions). Implementations of SCAI in a public network will be based on the public network switches, the same switches that may also support IN functionality (see Fig. 5.2).

If both IN and SCAI are implemented in the same switching systems and both are concerned with call processing, compatibility (or alignment) between IN and SCAI is necessary. Two aspects of the need for this alignment should be mentioned. First, such an alignment is a way of minimizing the differences between the two software packages (one for IN, one for SCAI) that the same manufacturer would have to create.

* As in the previous note, but related to IN.

This is helpful to a manufacturer whose goal is to minimize development efforts, and helpful to the network providers who buy these switches, because of software cost savings. Second, IN and SCAI will interwork to support certain services. This interworking will be easier (cheaper) if two systems are designed using the same approach.

The result of this principle is the development of the SCAI call model which is aligned with the IN call model. Functionality of the SCAI telecommunications-related services is defined based on the call processing described by this call model.

5.2.5 Usage in existing and developing networks

While new network concepts are being deployed (ISDN) or developed (IN), there are also existing networks (POTS). The SCAI standard, it was stated, should be applicable to existing or developing networks.

5.2.6 Security

This is an odd principle because it sounds more like a function than a principle, and it is included in the SCAI standard as a function. The reason it is also mentioned as a separate principle lies in the fact that SCAI is an external interface to a network and, therefore, requires specific security provisions without which it can't be deployed. This principle, identified early on, was a constant reminder of the importance of security for the SCAI as a standard.

5.3 SCAI Architecture

This chapter covers issues that can be placed in the architectural category. I use the architectural category to address the issues that have global significance for a particular problem: to outline how *the system* is laid out, what the components of this particular system are, and how they interact with each other—in general, how things fit together. This information should be useful for the design of functions, services, and the development of the system's evolution. There are three aspects of system architecture that are applicable: *functional, physical,* and *protocol.*

5.3.1 Functional architecture

The functional SCAI architecture identifies SCAI functional components, their main roles, and their relationship.

In Fig. 5.3, A represents a telephone device and B represents a data device (terminal) that a user is equipped with. A user in this functional architecture is someone (or something) operating both telephony and computing terminals. A user may be equipped with only one of those devices. While most functions in the current SCAI standard are geared toward a human user who is physically present, the SCAI architecture also allows for a more ambiguous user. A user is always an entity that benefits from the exchange of information across a SCAI link.

The switching function is a functional component representing the telecommunications environment. The computing function is a functional component representing the data processing (computing) environment. SCAI is a link between the two environments. SCAI performs the functions of coordinating the processes in the telecommunications and data processing environments for the benefit of a user.

Connecting lines shown in Fig. 5.3 represent logical connections. How the terminals are physically connected to switches and computers is part of the discussion of the physical architecture.

5.3.2 Physical architecture

Physical architecture defines how the various physical entities are connected to perform the required functions. The SCAI physical entities are switches, computers (hosts), and terminals. The SCAI physical architecture can be divided into three parts.

Topology. As a link between two environments, SCAI can connect single entities located in each environment (point-to-point configuration) or a combination of multiple entities located in each environment (point-to-multipoint or multipoint-to-multipoint). Figure 5.4 shows examples of various topologies. Although full networking capabilities

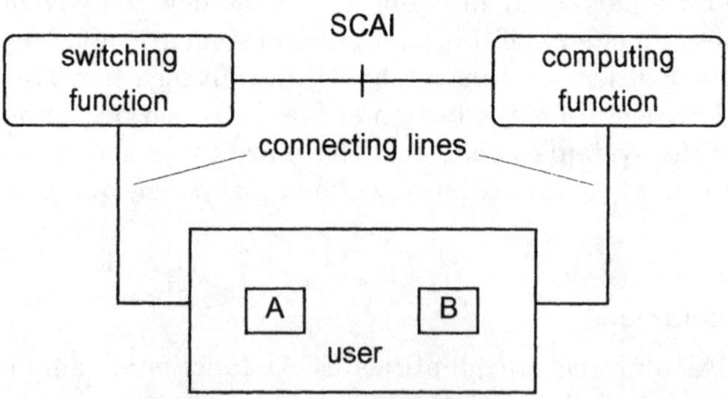

A - telephone device; B - data device

Figure 5.3 SCAI functional architecture.

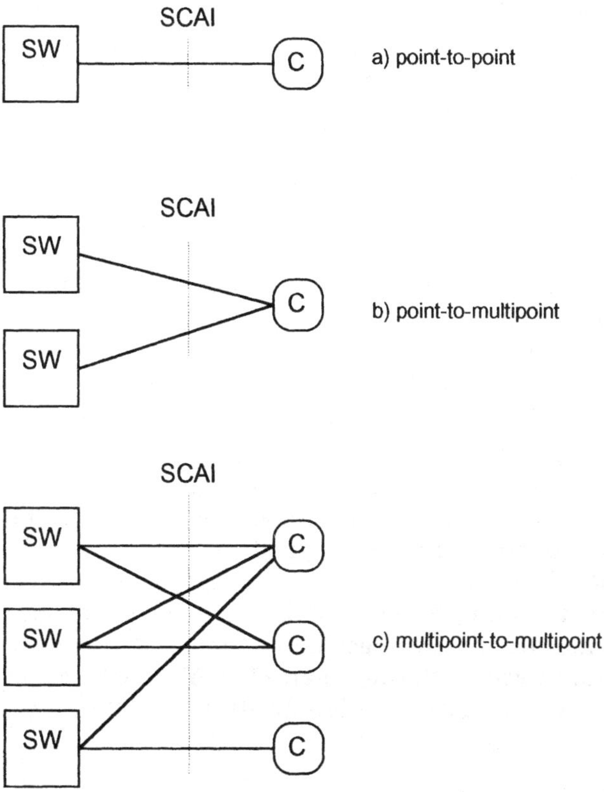

Figure 5.4 SCAI topology.

for SCAI are envisioned in the future, the current standard deals only with a simpler option—point-to-point. In order to extend the current SCAI functions to operate in a networking arrangement,* some additions would have to be made. The two most obvious areas for additions are addressing and information synchronization. Addressing would be necessary in order to find a path through the SCAI network to a proper destination (right now it is not required because communication is always point-to-point). Synchronization of information would be required if the same information is stored on various entities, and this will be the case if multiple computers are to be involved (right now this is not required because currently a SCAI interface is expected to connect a single computer to a single switch).

* Networking arrangement here means connectivity of more than two SCAI entities (switches or computers) to constitute a SCAI network.

Configurations. By "configurations" I mean a placement of switches and computers in public/private networks. There are several alternatives for switch-computer configurations (see Fig. 5.5):

1. Public network switch to customer premises computer
2. Public network switch to public network computer
3. Private network switch (PBX) to customer premises computer

Alternative 2, while possible, is not currently considered to be of great interest in the marketplace. Alternatives 1 and 3 represent configurations for which the standard was geared. These are the same two types of network that we already talked about, public and private.

Terminal connectivity. SCAI, in its most general concept, does not include the need for terminals. The concept is to merge telephone and computer environments for the benefit of the end users. However, the first phase of the SCAI standard was defined to support specific applications, all of which do expect users to be equipped with terminals. While the SCAI standard does not mandate how the telephones and computer terminals are to be connected to switches and computers, this is an important topic for the designers of SCAI networks and applications. There are two issues associated with terminal connectivity:

1. How terminals are to be physically connected
2. What types of terminals can be used

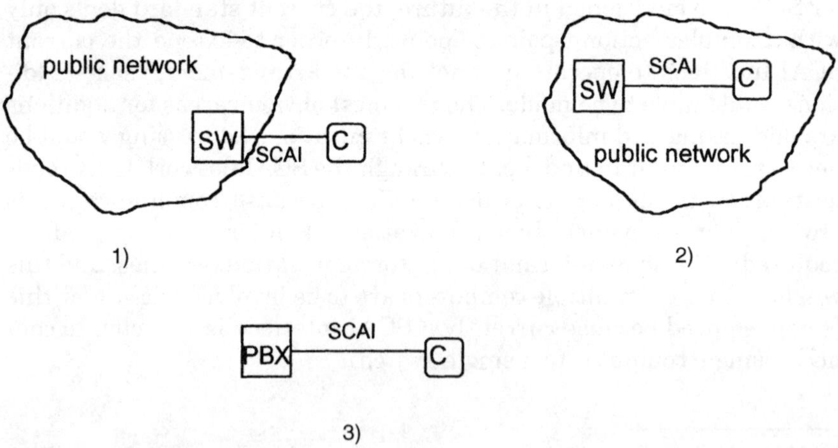

Figure 5.5 SCAI configurations.

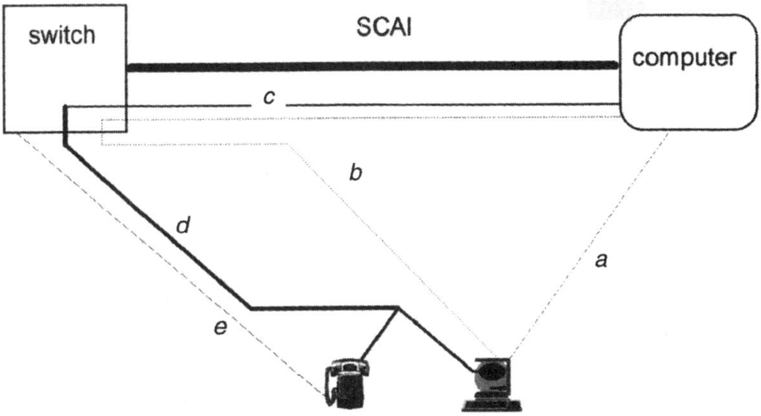

Figure 5.6 Terminal connectivity.

Figure 5.6 shows various alternatives for connecting computer terminals to computers. The telephone devices are usually connected directly to a local switch*; computer connections vary.

Alternative *a* is any type of direct connection of a terminal to a computer or a computer network, for example, LAN, MAN, a link in a star configuration using any appropriate protocol, and so on. In this case, a switch is not involved with the physical connection of a computer terminal, and a telephone is connected via interface *e*. Alternative *b* is a dedicated or switched connection of a computer terminal through a switch. It can be a WAN or a MAN connection, or a direct link to a computer using a leased line from a public network, or a case in which a PBX performs the functions of a LAN. The distinctive feature of this connection is that a switch is involved with the physical connection of a computer terminal. A telephone is connected via interface *e*. Alternative *c/d* consists of two parts: *c* is the connection of a computer to a switch using any kind of connection that suits the particular situation, such as a T1 interface. Portion *d* is an ISDN BRI which connects both a telephone and a computer terminal to a switch. In the ISDN case, a switch is involved with the physical connection of a computer terminal.

Figure 5.6 shows three different connecting lines between a switch and a computer: a SCAI link, option *b,* and option *c*. All of these con-

* The SCAI standard does not limit the connectivity of telephones to a local switch. From the SCAI architecture point of view, the connection of a telephone to a remote switch is possible, where the remote switch is a switch other than the one linked with the SCAI computer. The problem lies with the signaling that is required between a local and a remote switch. If such signaling exists, a remote connection may be possible; if such signaling does not exist, a remote connection is not possible.

nections can, in turn, be either on the same physical medium (e.g., T1) or not. This is a decision based on local conditions, traffic, and the availability of particular equipment. In real life it is possible to have all these different types of physical connections used in a single arrangement. Figure 5.7 contains computer terminals of three different types. Type 1 is connected through a dedicated line physically going through a switch. Type 2 is connected through a LAN. Type 3 is connected through an ISDN line to a switch and then through some kind of a data link to a computer. It is possible that the same switch and computer pair linked by SCAI supports users that are ISDN subscribers and POTS subscribers; in addition, the users could physically be located in different buildings and, because of this, connected to a computer through different means. In the example in Fig. 5.7, a SCAI switch and computer are physically connected by a T1 interface,* a widely used transmission system with specific characteristics that provide 24 individual digital channels† with bandwidth of 64 kbps each. Each individual channel is called DS0, while all T1 channels are referred to as DS1. One of the DS0 channels is used in this example for SCAI traffic: messages are transferred between a switch and a computer for SCAI purposes. Other DS0s are used for other purposes, including connectivity of computer terminals through the switch to the computer. In this case, the connection may be either dedicated or switched. The actual implementation of terminal connectivity depends on traffic volume, local requirements, and economics. Figures 5.6 and 5.7 contain two specific terminals: a telephone and a computer terminal that accesses information in a computer. But there can be different terminals and different combinations of terminals deployed, depending on the needs in a particular service environment. The telephones can be either 2500 sets, ISDN terminals, or headsets. The computer terminals can be "dumb" terminals or PCs. Users may have any combination of these terminals, and their SCAI actions would depend on what kinds of terminals they have. If the user has both a telephone and a computer terminal, commands to clear/transfer/conference calls can be issued from either terminal by performing specific actions (e.g., depressing a particular button). If, instead of a telephone, a headset is used, then the commands must be invoked on a computer terminal. If no computer terminal is needed for a particular service, all commands can be invoked on a telephone.

* Do not confuse a T1 interface with Committee T1; they are obviously different things. The context in which T1 is referenced should be sufficient for understanding whether a committee or a transmission system is meant.

† The T1 interface consists of 24 channels in the United States; in Europe it consists of 32 channels.

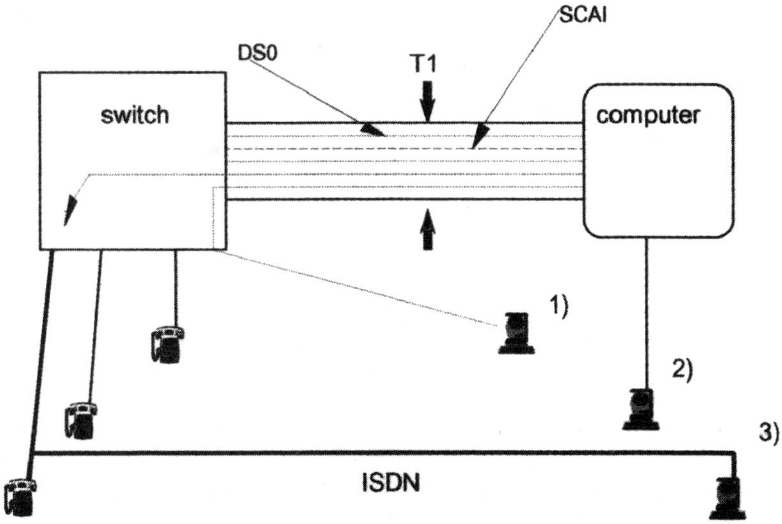

Figure 5.7 Example of physical connectivity.

5.3.3 Protocol architecture

The SCAI protocol architecture defines protocol components that are necessary to carry out the tasks of SCAI and their relationships. The SCAI protocol is an application layer protocol and to define SCAI is to define one or more ASE(s) that would constitute the protocol. In the current SCAI standard, two ASEs have been defined—SCAI and Security,* and these ASEs contain all functions deemed to be necessary in the first phase of the SCAI standard. In the future, the ASEs may be expanded to include more functions, or other ASEs may be created.

Part of the protocol architecture is the relationship of the SCAI ASEs with other protocol components that are necessary in the SCAI environment. We already discussed an issue of using other existing (standardized) functions or functional sets, if such are available, instead of defining the same functions again. Two such protocol components are included in the SCAI protocol architecture: ACSE and ROSE. ACSE† is designed to take care of association establishment and release; ROSE‡

* In the SCAI standard, T1.626, the initial architectural figure showing layer 7 ASEs contains the SCAI ASE and does not contain the SCAI Security ASE. A similar figure closer to the end of the document shows both of them. The reason for this is the desire to keep the discussion of the SCAI ASE separate from the Security ASE for clarity. In addition, the SCAI standard does not require the use of both ASEs; this decision is up to the service provider. More on this issue is in Secs. 5.6 and 5.7.

† See Sec. 4.3.

‡ See Sec. 4.4.

is designed to take care of interactions between remote entities. The resulting protocol architecture is shown in Fig. 5.8.

The SCAI ASE interacts with ROSE and SACF. ROSE and ACSE interacts with the presentation layer. The SCAI Security ASE interacts with ROSE. SACF provides coordination between the ASEs. Figure 5.8 is a part of the standard; what it does not show is what is going on above SACF. This was done on purpose as the result of many standard development participants' objections to standardizing anything that resembles applications or services. Most participants (and this is true for other standards development beside SCAI) are not willing to standardize what they consider to be part of their competitive advantage. In reality, though, we can redraw this figure to indicate more precisely what protocol architecture exists in an implemented system (Fig. 5.9).

Interface *a* is an interface to the presentation layer. Interface *b* identifies the relationship between ACSE and SACF rules. It is invoked when there is a need to use ACSE, for example, when an established association is to be released. Interface *c* is an interface between SCAI ASE and ROSE and is used by the SCAI ASE to send/receive information using an established association. Interface *d*, while not formally defined by the SCAI standard, is used by the SACF to access the SCAI ASE. This interface is not formally (explicitly) defined in the standard because it was seen as an implementation-specific issue. This is a simi-

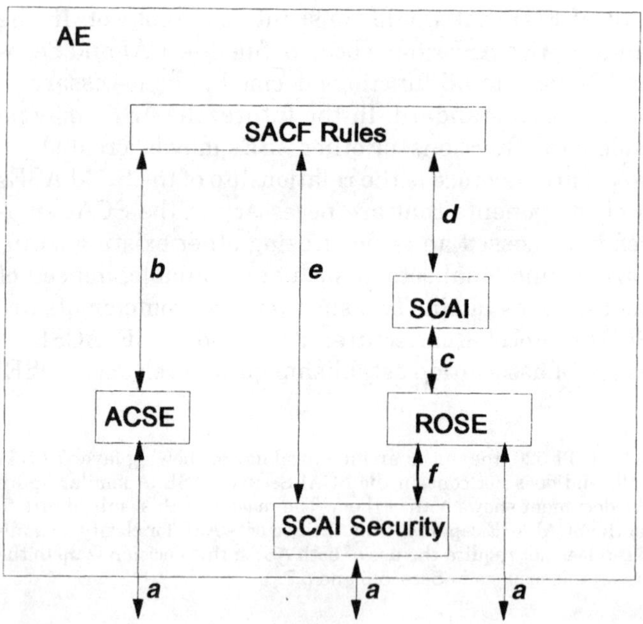

Figure 5.8 SCAI protocol architecture.

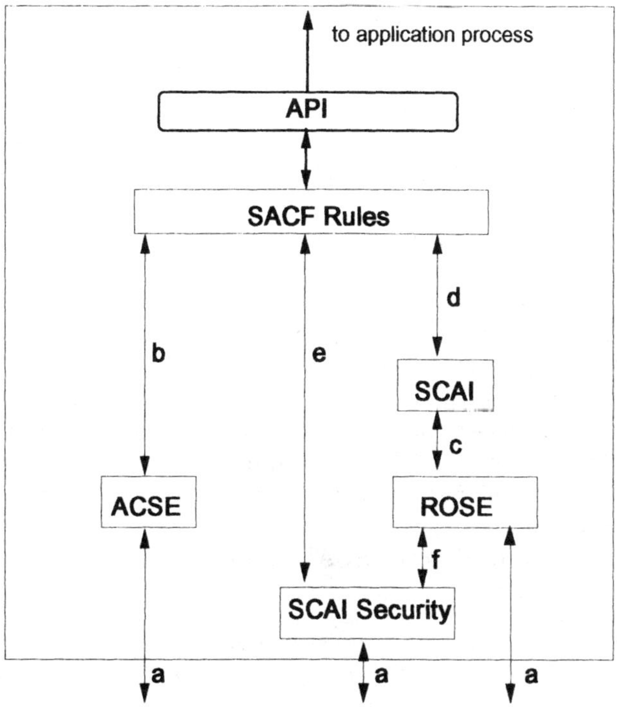

Figure 5.9 Protocol architecture with API.

lar issue to the API; some participants felt uncomfortable standardizing anything approaching an interface with application processes. Interface *e* is used by the SACF to access the SCAI Security ASE. Interface *f* indicates the flow of information from ROSE to the SCAI Security ASE, whose function is to encapsulate ROSE PDUs in secure envelopes.

The SCAI standard defines:

- SCAI ASE
- SCAI Security ASE
- SCAI ASE's interactions with ROSE
- SACF rules
- Use of ACSE

5.4 SCAI Modeling

Modeling is the technique of describing the behavior of a system or a unit in general terms. Four models are used in the SCAI standard: modeling of objects that are visible in SCAI, modeling of call states, modeling of party conditions, and modeling of agent states. Together they constitute the SCAI switching model, which is defined in the T1.626 standard as ". . . an abstract view of switching objects and their behavior." The SCAI call model, party model, and agent model are finite state machines, or dynamic models, representing different states and state transitions. The SCAI object model is a static description of objects and their relationships.

The development of SCAI began just prior to the time when an object-oriented approach to modeling and design became popular in the standards community. While SCAI developers did not use this technique fully, some elements of it were applied. The emphases were on understanding the properties, behavior, and relationships of objects perceived to be interesting to SCAI applications. These objects were chosen based on practical experience with the applications and equipment supporting CTI functionality. Understanding of the SCAI objects provided the standard's developers with the knowledge necessary for the creation of the protocol.

There are four objects defined in the SCAI switching model:

1. Call
2. Device
3. Party
4. Agent

The significance of the objects that are defined is that they are "visible" to an application process that uses SCAI, and, therefore, this application process can address and manipulate them. For example, "visibility" of a call object implies the ability of an application to make decisions regarding a particular call: whether to proceed with it or clear it, how to terminate it, and so on. Although the SCAI standard does not formally define it, there is the underlying concept of a "visibility domain" that supports this notion. Let's consider the following example.

Five hundred customer service agents are located at five hundred devices (telephones) and served by a switch. The switch is linked with a SCAI computer running some SCAI applications (services). Four hundred devices are identified to the application as the devices to be served by the application (customer service center). One hundred devices are excluded from this relationship and are used for other purposes. We can then say that 400 devices are in the SCAI domain. These devices, by name (number), are visible to the application which can "control" them and "observe" the activities at them, if it desires. A specific action is required for the device from the group of 100 to be placed in the SCAI domain. The action can be static (performed when the

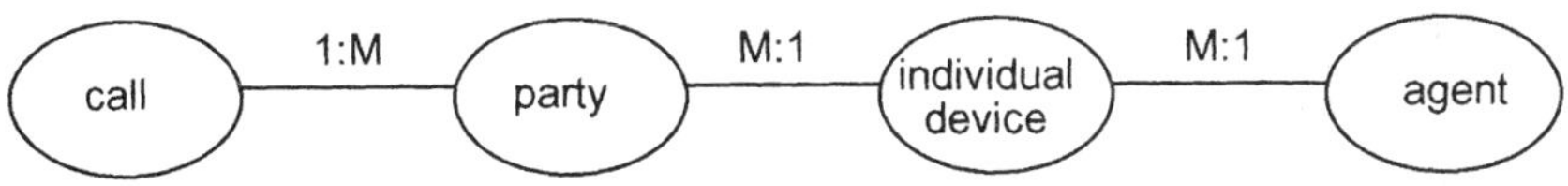

Figure 5.10 SCAI objects relationship.

entire arrangement is reconfigured) or dynamic (performed by an exchange of messages between a switch and a computer, while the entire SCAI arrangement is operational*).

All SCAI objects must be in the SCAI domain (visible to an application) in order to be a part of the SCAI operations. All SCAI models are discussed in the rest of this chapter, but first let's look at Fig. 5.10 showing the relationships of the SCAI objects.

A question may be asked regarding Fig. 5.10: "Why is an object class *device* represented in the figure by one of its subclasses, *individual device?*" The reason is that a subclass *group device* has a behavior different from that of an *individual device* (see Sec. 5.4.2).

Table 5.1 summarizes the meaning of each relationship.

5.4.1 Call object

A call object is defined in the T1.626 standard as ". . . a switching function communications relationship between parties." The number of parties participating in a call may range from 1 to *n,* where *n* equals as many parties as can be serviced by particular equipment or arrangement. A call may involve only a single party in certain circumstances, for example during a call setup.

A call class has two subclasses, as shown in Fig. 5.11. Let's consider two subclasses of a class *Call* independently.

Two-party call object. A two-party call is a call involving up to two parties. It has the following attributes:

1. *Call identifier (call ID).* It uniquely identifies the two-party call. MEANING: Each call that a switch is manipulating (establishes, releases, transfers, etc.) is assigned an ID. This ID is used to track the call as it moves through various stages and involves various participants. In SCAI, the call ID uniquely identifies the call at a switch. However, the SCAI standard does not dictate how the ID assignment

* This dynamic ability is considered SCAI management and is not currently covered by the SCAI standard.

TABLE 5.1 SCAI Object Relationships

Relationship	Meaning
One-to-many relationship of a ***call*** to a ***party***	A call can have many parties participating in it, and a party can be a member of only one call
Many-to-one relationship of a ***party*** to an ***individual device***	An individual device may have many parties at this device, but a party can be associated only with one individual device
Many-to-one relationship an ***individual device*** to an ***agent***	An individual device may have an agent signed on, and an agent may sign on at many individual devices

and management is accomplished. For example, in one switch implementation, a call, when transferred, would be assigned a new call ID, while in another, the call ID would not change. Sometimes, the decision on how to track calls is also based on the needs of a particular application. If the user's application requirement dictates that during a transfer the call must retain its original call ID, then an implementation must be designed with that requirement in mind. Again, SCAI does not impose any restrictions in this regard.

2. *Call state.* It identifies the state of a call.
   MEANING: Each call transitions through states (see Sec. 5.4.5 for more details). This attribute identifies the state of a call and the information is used by the application process to make decisions regarding its progress.
3. *Parties.* It identifies participants in a call.
   MEANING: Party is an abstraction. For more details see the discussion of the party object.

Conference call object. A conference call is a call involving more than two parties. It has the following attributes:

1. *Call ID.* It uniquely identifies the conference call. Assignment and management of conference call IDs is not specified by the SCAI standard.

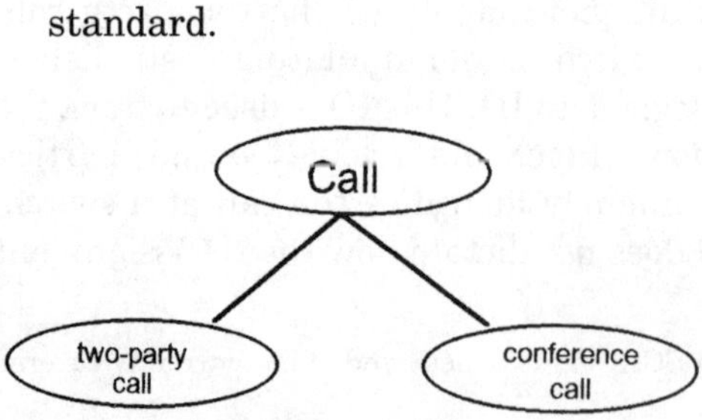

Figure 5.11 Call class.

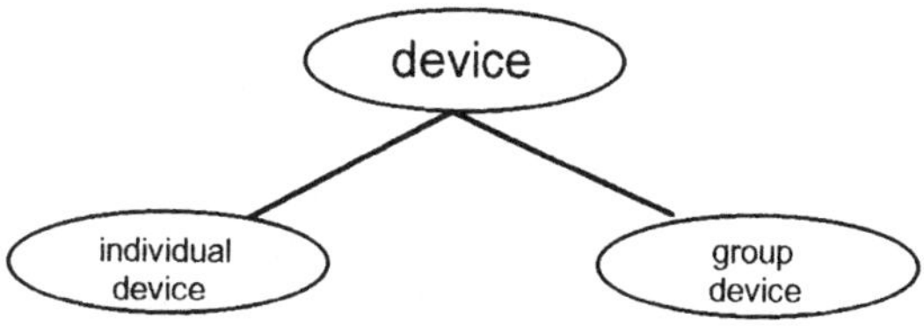

Figure 5.12 Device class.

2. *Parties.* Participants in the conference (see discussion of the party object for more details).
3. *Conference condition.* It identifies a condition* of the conference. MEANING: Conference conditions identify specific states of a conference call, which are:
   - conference null—no conference exists
   - active—at least one participant is involved with a conference
   - float—a conference is active, but without a controller†

5.4.2 Device object

A device is defined in the T1.626 standard as "... an identifiable entity which can participate in a call." The easiest way to explain a SCAI device object is to imagine a physical device that is the destination for the call. These devices can be of two types (as in Fig. 5.12): individual and group devices.

Individual device object. An individual device is an identifiable entity that has the appearance of a single line. The most obvious device that fits this definition is a single-line telephone. Individual devices are the terminals that are used by human operators, but we can also imagine them being automatic devices that work under the control of computers. The individual device object has the following attributes:

1. *Device type.* It identifies the particular type of device:
   - single-line appearance on a single-line station
   - single-line appearance on a multiline station (e.g., ISDN)
   - headset (no keypad)

---

* Actually, these so-called conditions are states of the state machine that describes behavior of a conference. But, because the SCAI call model is described in terms of states, the decision was made to use a different term in the case of a conference call object (and in the case of a party object). The states of a call model are visible to a much larger set of functions than the conditions of a conference call.

† A conference controller is a party that is in control of the conference call.

2. *Device identifier (ID).* It uniquely identifies an individual device. MEANING: Device IDs are assigned by the switch for the purpose of addressing the devices. Since addressing can be accomplished through the use of different types of numbering schemes, SCAI allows a great variety of them. A device ID can be, for example, a public directory number, or a special number used internally by a switch to address the devices connected to it.

3. *Parties.* Each party associates a device with a call (see party object).

4. *Agent identifier (ID).* It identifies an agent at a device.

Group device object. A device group is defined in the T1.626 standard as "... an entity that distributes calls to devices." Devices to which calls are distributed are members of a group and can be either individual or group devices, as shown in Fig. 5.13.

When calls are addressed to a device group, it really means that they are addressed to one of the members of the device group, and the group has the responsibility to deliver a call to an appropriate destination (its member). How the distribution is accomplished is outside the scope of SCAI. We can examine an example of the device group: ACD. *Automatic call distributors (ACDs)* are responsible for the distribution of calls to the first available agent, or to a specialized agent, or to a specific (requested) agent. An ACD consists of queues for all possible destinations, and, when a call arrives at an ACD, it is placed in one of the queues based on some criteria. For example, calls originated from one geographical location are placed in a queue to a specific set of agents. Another example is the placement of a call in a queue for a specific agent after a caller entered the name of that agent.

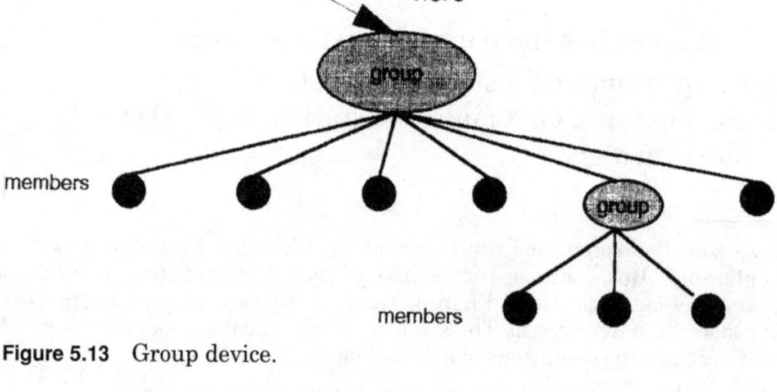

Figure 5.13 Group device.

The attributes of a group device object are:

1. *Group type.* It identifies the type of a group. Examples are:
   - ACD group
   - Hunt group
2. *Group ID.* It identifies a group; this ID takes on the same form as the ID of an individual device.
3. *Group members.* It identifies members of the group; they can be individual or group devices (see Fig. 5.13).
4. *Calls.* It identifies calls that have arrived at the group device and have not yet been distributed.

5.4.3 Party object

A party is defined in the T1.626 standard as ". . . an identifiable participant of a call." This is an associative object that represents the relationship between a device and a call. The party object is necessary because in certain situations a state of the call is not a sufficient description of what is actually going on with this call. An example is a situation in which a call is placed on hold. After this action, the call is still in the same state (according to the SCAI call model, described further) as before, but actually something else is now different: the call is on hold. Of course, this could be modeled as another call state, but the party object is useful in this situation because it allows a call to be modeled as still in the same state, and a party associated with this call is on hold.

The party object has the following attributes:

1. *Party identifier (party ID).* It uniquely identifies a party.
   Meaning: Party ID is a reference of a specific participant in a call. It is a combination of a call ID and a device ID. Both call ID and device ID may be provided or either one can be provided, depending on the information available in a specific instance.
2. *Party condition.* It identifies the condition* of the party (see the following).
3. *Call ID.* It identifies a call (see *call object*).
4. *Device ID.* It identifies a device (see *device object*).

---

* As in the case of conference conditions, party conditions are states of the finite state machine (see *conference object*).

Party conditions. Three party conditions (states) are defined. The party condition model is shown in Fig. 5.14.

When a party is not involved in a call, it is said to be in a NULL condition. When a party is involved in a call, it is in an ACTIVE condition. When a party is placed on hold, it is in a HOLD condition. From a HOLD condition, a party transitions either to a NULL condition or an ACTIVE condition (if the call is retrieved).

One point is important to understand about the party object (because it affects the understanding of some functions described later): when an instance of a party object is referenced, it is identified in terms of a combination of device and call IDs. Either one or both of them can identify a party. The change of either ID manifests itself in the change of the party identification. For example, imagine a party identified as party *XY* constructed of device ID *X* and call ID *Y*. Let's say that a second call is coming to the same user at the same device *X*. If the user wants to answer the new call, the current call has to be either placed on hold or disconnected. The new call then would have a different call ID, let's say *Z*, and, although the device ID has not changed, a new party is created. This party is *XZ*. The same user at the same device, therefore, can be associated with different parties.

5.4.4 Agent object

Agent is a different type of an object in comparison with other SCAI objects. First, the agent object represents a real human being. There is no abstraction involved; an agent is there because a particular application (e.g., customer service center) requires a human being to answer the phone. Second, the agent object is included in the SCAI object model for a different reason than other objects. While it is important for SCAI to keep track of calls, devices, and parties (and this is the reason these objects are described), agents do not play a real role in the technology. For example, if a particular agent (by name) has to take particular calls, SCAI does not have to know the name of that agent; instead, a device at which the agent is working is used as the destination address. So, why is the agent object described in SCAI? The reason

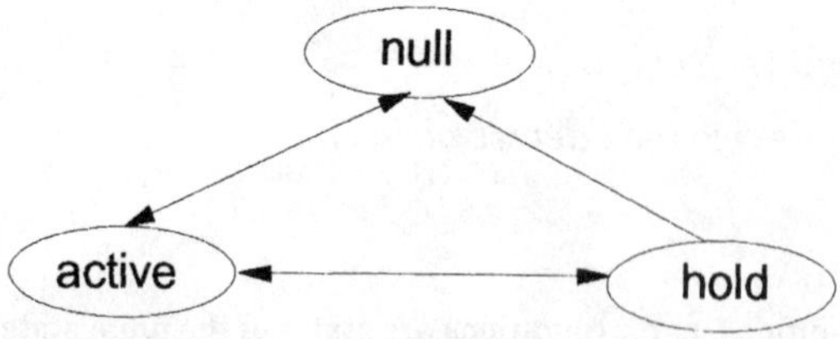

Figure 5.14 Party conditions model.

lies in the other need of an application: the need for statistics! It is important for applications dealing with real human beings (agents) to collect statistics on the available resources, productivity (e.g., how fast a particular agent serves a customer), peaks for specific agents, and so on. This information can be used later by management either to make changes to increase productivity, or to dynamically affect resource allocation. The agent model, which we will consider soon, reflects the states of an agent regarding the agent's working status.* In the future, features controlling some agent operations will also be defined. But this version of the SCAI standard does not contain these features. An agent object has the following attributes:

1. *Agent identifier (agent ID).* It uniquely identifies an agent. MEANING: When an agent signs on the device, this signature is recorded as a name or a reference that would distinguish this particular agent from others.
2. *Agent password.* A typical computer access security device; it allows a system to recognize the agent.
3. *Agent group identifier.* It identifies the group to which the agent belongs. An agent can belong to many groups (this capability is allowed by SCAI, but in real life this ability is controlled by the particular circumstances of the operating environment).
4. *Agent state.* It identifies the state of the agent (see the discussion on *agent model*).
5. *Device identifier.* It identifies the device at which the agent signed on (see *device object* for more details).

Agent model. The agent model (Fig. 5.15) is used to identify the state and state transitions of an agent.

The states are defined as follows:

NULL—the agent is not signed on any device; this agent is not participating in any activity.

---

* It is interesting to trace this seemingly obscure technical issue to a more socially visible concept of privacy. Here, technology helps to diminish the privacy of a worker, an agent, by keeping the following statistics on all aspects of a worker's workday: how many minutes a worker spends on a customer, how often a break is taken, how long a pause exists between two successive customers, etc. This definitely helps to increase productivity, but does privacy suffer? With computers spreading through all aspects of our lives, this question is being asked more and more often. Sooner or later there will be a political debate on this issue, but in the meantime, technological solutions like SCAI are mostly interested in the economic benefits derived from technology. What are the engineers to do?

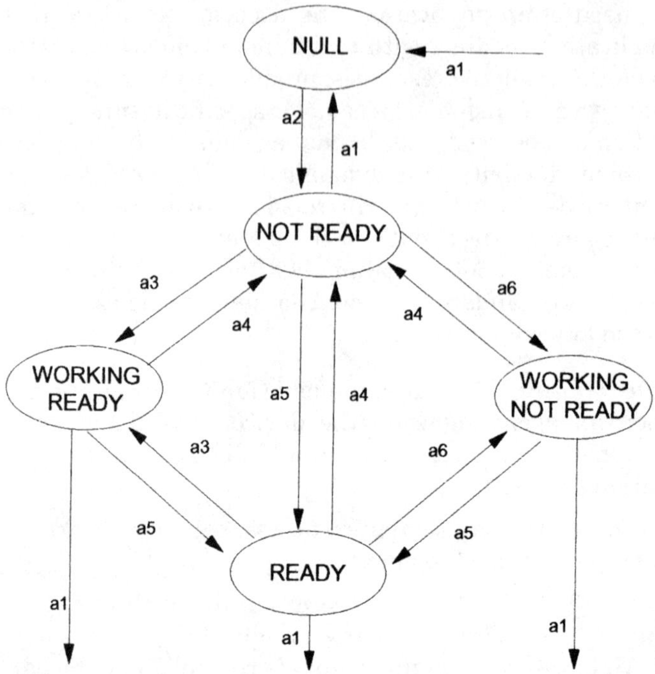

Figure 5.15 Agent state model.

READY—the agent is signed on a device and is ready to receive calls.

NOT READY—the agent is signed on a device, but is not ready to accept calls (e.g., the agent is on a five-minute break).

WORKING READY—the agent has finished the previous call with a customer and is currently involved with the processing of information related to that call; the agent may receive other calls (e.g., although the agent is still busy, the business conditions dictate the need to serve a new customer).

WORKING NOT READY—the agent has finished a previous call with a customer and is currently processing information related to that call; the agent cannot receive other calls (the agent cannot serve another customer under any circumstances).

Agents can receive calls when they are in READY and WORKING READY states. WORKING READY and WORKING NOT READY states are similar: they both imply that the agent is still occupied with unfinished business from the previous call. The difference between them is that, in the former case, the agent can receive calls, and, in the latter case, the agent cannot receive calls. The difference is based on two approaches by the makers of call distributors to the treatment of the agents that are finishing the previous call. It is envisioned that a call distributor (a switch, in SCAI terms) models agents with one of these two states.

State transitions are reported by the event reports shown in Table 5.2.

5.4.5 Call model

The SCAI call model is a model of a basic two-party call. It describes the states and transitions of a two-party call. It describes a basic call only; features that are additional to the basic call are described separately. The examples of these features are call transfer, call hold, call retrieve, and so on. The decision not to include features (or, using ISDN terminology, supplementary services) in the call model was made because of the desire for a simple, easy-to-describe and -understand model that is applicable to all calls. Supplementary services may be different for different arrangements and, in addition, there will be more of them in the future. If one would attempt to include all state transitions for all supplementary services, the diagram of the call model would become unintelligible.

The basic SCAI call model is a state machine showing the states, transitions, and the reason for transitions. The call model is a common thread for all calls. It is one of the most basic building blocks in SCAI because all telecommunications services are based on the call model. The key concept that is necessary to understand about the call model is its view of a call, which is a single-ended view. Let's look at this concept in more detail.

When a call is being made, a telephone network is responsible for setting up a connection between involved parties (or, in telephony terms, a call is being established). If we simplify the process to a single network switch, we can identify the stages of this process more easily.

When party A calls party B (as in Fig. 5.16), a switch references this call attempt using a specific scheme implemented in a switch. There are two basic alternatives in referencing: half-calls and full calls. Other terms for these alternatives are, respectively, a single-ended view and a global view of a call.

TABLE 5.2 Agent State Transitions

Label	Event report
a1	Logged_Off
a2	Logged_On
a3	Working_Ready
a4	Not_Ready
a5	Ready
a6	Working_Not_Ready

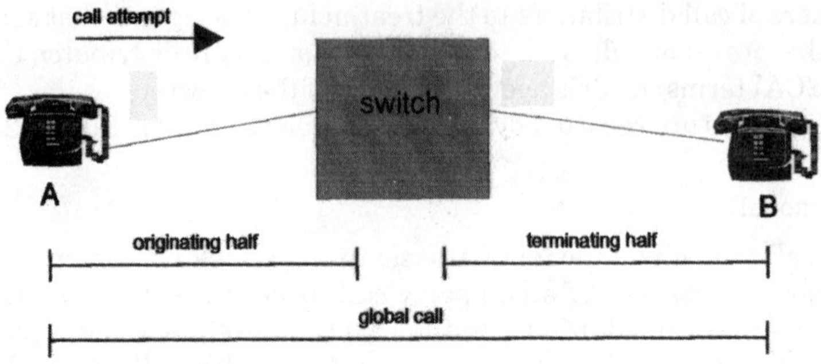

Figure 5.16 Single-switch call.

In the case of a single-ended view, our call from A to B would be considered as two halves: originating and terminating. When A initiates a call by dialing a number, a switch receives the dialed digits and assigns a reference* for this *originating* half of the call. Then a switch proceeds to establish a call to B and assigns a reference for this *terminating* half of the call. The result is that what is perceived to be a single call is actually treated by a switch as two halves of this call.

In the case of a global view our call from A to B would be considered a single call and assigned a single reference. So, why bother with a single-ended view if a global view seems to be simpler? The reason is that most of the calls are originated and terminated at different switches (see Fig. 5.17).

In most situations, when A calls B, switch #1 receives the request, assigns this call attempt a reference, and passes it on to the next switch. Switches #2 and #3 do the same thing and, finally, switch #4 terminates the call at B. Switch #1 does not know which switch will

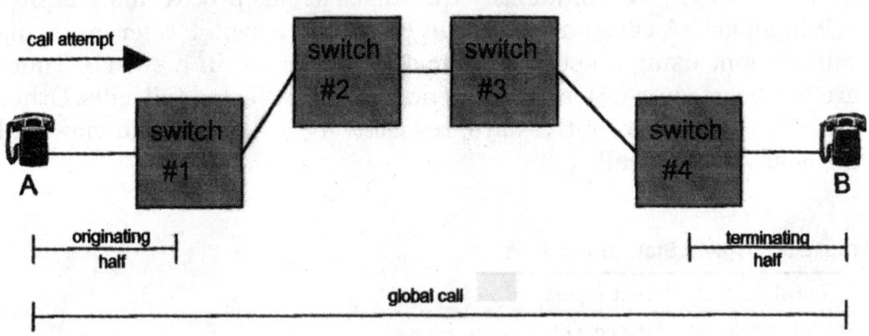

Figure 5.17 Multiswitch call.

* A reference is a value assigned by a switch to a particular object instance, in this case a call, to keep track of its status. In SCAI terms, this reference is a call ID.

terminate the call, or whether the call will be terminated at all. But it does know how to reference what it sees to be an originating call from A. Switch #4, on the other hand, knows how to reference what it sees to be a terminating call to B.

If we now look at two alternatives it is easy to see that in the case of a single-ended view the situation is simple: each switch controls a specific half of the call. In the case of a global view, things are more complex. Both switches, #1 and #4, would have to be aware of both halves of the call: who and when originated the call, where it is terminated, and so forth. There are two problems with this approach. First, appropriate signaling is required to carry all necessary information between the switches. While SS7 is capable of fulfilling this requirement, not all networks are fully equipped with SS7 at this time. Second, synchronization of call references is required and it can be very complex, especially in the cases of switch failures.

Since most calls are not handled by a single switch, the single-ended view of a call is used in SCAI. According to this approach, the call model is split into an originating half and a terminating half. The principles of modeling are, of course, the same for both halves, but the states are different. Each half-model includes call states, transitions between states, causes for transitions, and reports associated with transitions. Originating and terminating call models are shown in Figs. 5.18 and 5.19. Each transition arch is marked by a label that is explained in Tables 5.3 and 5.4. Each label corresponds to a specific event report that is issued when a call transitions from one state to another.

Table 5.4 has one inconsistent entry—the last one. This is done for the visibility of this transition because it is different from all other transitions in the call model. This transition does not generate an event report. Route Request is not an event report, but a request to perform an action—to find an appropriate route. To be consistent with the rest of the model, the transition from NULL to ROUTING should generate a report, for example, Route Requested, and then a request to find a route would be generated within the ROUTING state (see Fig. 5.20).

Although this solution is good for consistency, it creates another protocol element (another report) that is not really necessary because the reception of Route Request serves not only as an action request, but also as an indication of the specific transition. Route Request is actually shown in the call model of the SCAI standard with the brackets around it to focus attention on this aspect of the call model.

Call states. The states of the SCAI call model represent call progress for each call. Call progress is a reflection of call processing that takes

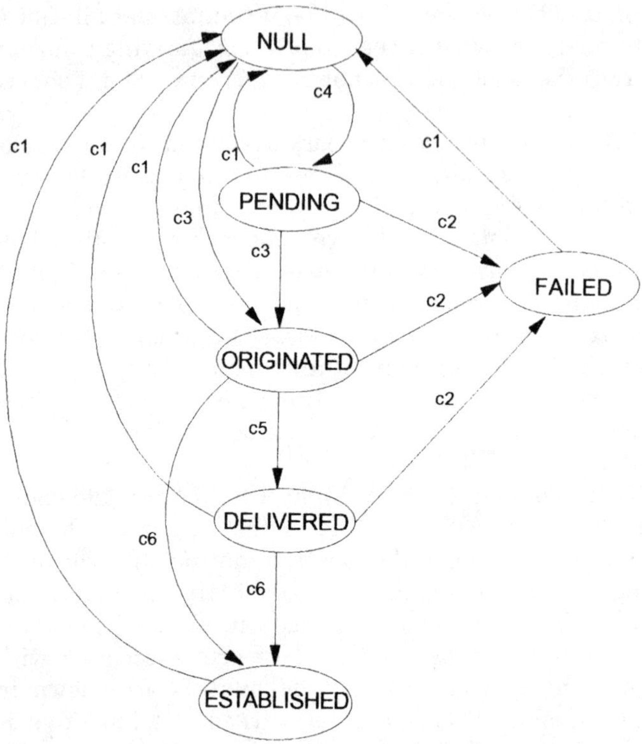

Figure 5.18 Originating call model.

place in a switch. If we look at a call model of any switch designed by any manufacturer, we see a very detailed state machine that models every action during call progress. For example, collection of each dialed digit would be represented by a state transition. The SCAI call states are not as granular. There is no state in the SCAI call model that signifies a collection of a single dialed digit, instead, the SCAI call states are what can be called macro states: they consist of what would be in other call models several states. The reason for this lies in the difference of the knowledge that is to be derived from the call states and their tran-

TABLE 5.3 Originated Event Reports

Label	Event report
c1	Call Cleared
c2	Call Failed
c3	Call Originated
c4	Service Initiated
c5	Call Delivered
c6	Call Established

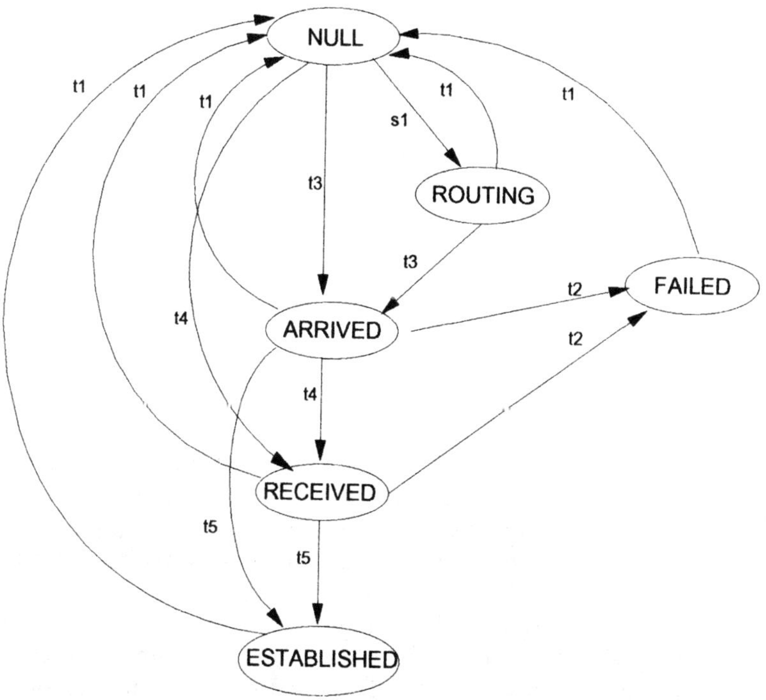

Figure 5.19 Terminating call model.

sitions in different applications. When an application like call processing in a switch is involved, it must be aware of each small action. Otherwise, the call processing application cannot manage call progress. When an application that uses SCAI is involved, such as call management, it is interested in knowing about some important events in the life of a call: all digits are collected or *all* information necessary to proceed with a call is collected. Other events would not be useful. The result of this reasoning is a much more compact and easily described call model.

TABLE 5.4 Terminating Event Reports

Label	Event report
t1	Call Cleared
t2	Call Failed
t3	Call Arrived
t4	Call Received
t5	Call Established
s1	[Route Request]

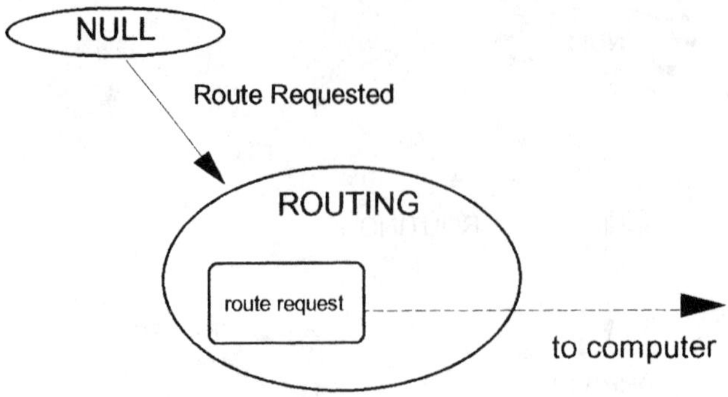

Figure 5.20 Possible modeling of ROUTING state.

Let's now look at each individual state of the call model. Each call state is described in terms of several characteristics collected in the form of a template. The template described in Table 5.5 was created through a process of trial and success. This is true not only for this particular template, but also for other templates used in the SCAI standard (and described in this book). When a new standard on a new subject is being defined, it is difficult to know up front what characteristics of one or another object will be of interest to an application or to a designer who uses the standard. The process of creating these templates was iterative. When a function was being defined and existing characteristics were not sufficient, the template was modified to include, change, or, sometimes, exclude some items.

Each state of the SCAI call model is described using the template in Table 5.5. Table 5.6 summarizes call states descriptions.

A word about the ROUTING state. It is unique in the SCAI call model. This is the only state that, instead of progressing a call, sus-

TABLE 5.5 Call State Template

Description	Defines the state
Entry event reported	Identifies the valid event that caused the transition into the state
Transition cause	Defines the valid reason(s) for leaving the call state
Originating device	Identifies transition causes for the originating device
Terminating device	Identifies transition causes for the terminating device
Request FE*	Identifies the valid request from the client that may cause a transition out of the state

* FE is a *functional element*—an "atomic" function which is a part of the SCAI ASE. More on this subject appears in Sec. 5.5.

TABLE 5.6 Call States

Call state	View	Description
NULL	Originating, terminating	No call exists
PENDING	Originating	A switch is collecting information necessary for identifying the destination
ORIGINATED	Originating	A switch is performing address analysis, route selection, authorization, etc., and initiates routing of the call to the destination
DELIVERED	Originating	Terminating end of the call is alerting (attempting to complete a call establishment process)
ROUTING	Terminating	It is a wait state; call processing is suspended while a switch awaits routing instructions from a computer
ARRIVED	Terminating	A call has arrived at a switch and the destination has been identified; call distribution, queuing, and hunting may be performed
RECEIVED	Terminating	A switch is alerting the terminating end of the call to complete the call establishment process; corresponds to the originating DELIVERED state
ESTABLISHED	Originating, terminating	Two ends of the call are connected; the call is established (complete)
FAILED	Originating, terminating	Normal progress of a call is aborted due to a discovered inconsistency or an error

pends the call processing. It is done in cases when a switch cannot route the call due to lack of information, or when an application process actually controls who will receive a call. This routing function is available only on the terminating side of the call because, so far, there was no application found to be interested in this function during origination, when routing is performed according to the existing network algorithms.

The SCAI standard contains a list of transition causes. Transition causes are indicators of the reasons for a particular transition. In line with the principle of close ISDN relationship, many of the transition causes are identical to ISDN causes defined in the ANSI standard T1.607 (circuit-switched portion of the ISDN access protocol, DSS1). Others are specifically defined in the SCAI standard.

As an example, let's consider the DELIVERED state described in Fig. 5.21.

Description The terminating end of the call is alerting.	
Entry Event Reported:	Call_Delivered
Transition Causes:	**to State**
Originating Device:	
■ No Answer	FAILED
■ Call Cleared	NULL
■ Host Initiated Clearing	NULL
Terminating Device:	
■ Destination Answered	ESTABLISHED
Request FE:	
■ Clear_Call	NULL

Figure 5.21 DELIVERED state description.

The description of this state is clear and there is no need to elaborate on it. The entry event reported when the call transitioned from the previous state is Call_Delivered. This means that during the transition the Call_Delivered event was sent to a computer; the expectation is that the computer's models are updated based on the receipt of this event. Transition causes for both originating and terminating device are listed, which means that the activity of either an originating or a terminating device can cause a transition out of the DELIVERED state into some other state. Specifically,

- If an originating device does not answer, the call transitions to the FAILED state
- If an originating device clears the call (an agent hung up the phone), the call transitions to the NULL state
- If a request is received from the SCAI computer on behalf of the originating device to clear the call, the call transitions to the NULL state
- If a terminating device answers, the call transitions to the ESTABLISHED state

A request FE (see Sec. 5.5) that can cause transition out of this state is, in this case, Clear_Call. When this request is received from a computer, a switch performs the requested function* and the call transitions to the NULL state.

We discussed that the SCAI call model consists of the originating and terminating halves. We need to recognize that most calls associated with SCAI are either originated or terminated by the switch directly involved with SCAI. In these cases, the switch tracks the progress of a specific originated call or a specific terminating call. For example, if a customer calls a service center, the switch that terminates the call (and which happened to be connected to a SCAI computer) "sees" only the terminating portion of that customer's call. As the call progresses, the switch tracks it based on the terminating call model. If a SCAI computer originates a call to a customer, the switch sees and tracks the originating portion of the call; the terminating portion of the call is handled by a remote terminating switch. There may be situations in which the entire call is handled by the same SCAI switch; for example, a customer who calls a service center happened to be serviced by the same switch that serves the customer service center's agents. In this case the switch sees both halves of the call and has to track the progress of both halves. How the switch does it is not important for SCAI. It is important, though, for a reader to understand that in this case the originating and terminating NULL states are the same, and the originating and terminating ESTABLISHED states are also the same. Before the call was originated, there was no call (NULL), and when the call was completed, both halves were in the same ESTABLISHED state.

In the works: The device object may be modified; a possible modification is envisioned for several reasons: (1) to differentiate between lines and trunks, (2) to clarify the role and behavior of a group device, and (3) to model a terminal (and possibly components of a terminal) as a separate subclass. These changes would allow SCAI to support more functions associated with all mentioned objects.

Transitions to the ROUTING state from states other than NULL may be added. This would mean that the routing request could be issued in more than one state.

5.5 SCAI Functionality

We already discussed the functional role of SCAI as a bridge between telecommunications and data processing environments. We also dis-

* A switch performs the requested function unless there are some set rules or restrictions prohibiting the switch from honoring the request, in which case an error response to the computer is generated.

cussed the SCAI architectural and modeling concepts based on which SCAI can become the bridging technology in a generic sense. Now it is time to discuss the specific functionality addressed in the SCAI standard. One thing to remember is that the first phase of the standard contains the functions that were deemed necessary in the today's marketplace and, at the same time, could be agreed to in a timely manner by the standards community. Future additions and enhancements to the standard should be relatively easy to do based on the current architecture.

From a functional point of view, SCAI has two important characteristics:

1. It is a coordination mechanism
2. It provides third-party call control functionality

The notion of a coordination mechanism is directly derived from the view of SCAI as a bridging technology between two environments. One of the most obvious functions for such a technology is coordination of processes in these two environments. In the chapter on applications it is described how the users benefit from this technology. In general terms, coordination implies an automatic exchange of information between two environments in a timely manner. Based on this exchange of information, decisions are made by a switch or by a computer that otherwise would require, if a particular service is to be supported, some special, nonautomatic, actions in both environments. These nonautomatic actions would be a lot less effective than the SCAI's provided coordination.

The notion of the third-party call control is based on the need to perform certain functions on behalf of a specific user. Figure 5.22 shows three parties*—A, B, and Z. In one scenario, party A (first party) makes a call to party B (second party). When a call is established both parties are involved in a normal two-party call. In another scenario, party Z (third party) issues a request for a call setup on behalf of party A, with party B being a destination for the call. When the call is setup, parties A and B maintain their connection for the duration of a call, but this call was established due to the actions of party Z and, therefore, constitutes a call generated by a third party. Hence, third-party call control is the ability to perform call control functions on behalf of another party.

Keeping in mind these two characteristics would enable a reader to understand the overall functional approach to the SCAI standard.

---

* These parties should not be confused with the formally defined SCAI object *Party*. In this description party is what it means in plain English—someone on a phone.

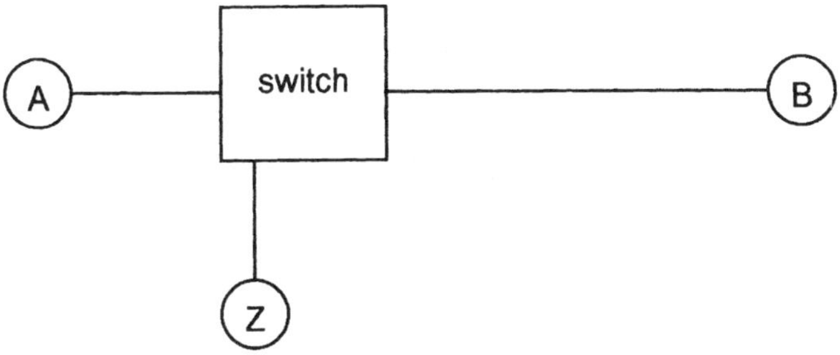

Figure 5.22 Third-party call control.

5.5.1 SCAI services

SCAI services are functions that perform specific actions. A SCAI individual service is called a *functional element* (FE). The SCAI ASE is a collection of FEs (see Fig. 5.23). All FEs defined in the SCAI standard could have been placed in any number of ASEs, but the decision was made that a single ASE is appropriate. From now on in this book, the term FE will be used to mean a SCAI service.

An FE is an atomic building block of the SCAI ASE. Each FE, when defined, constitutes a protocol element that carries out a specific service. Three classes of FEs are identified:

1. *Request.* An FE used to request that a function be performed
2. *Event.* An FE used to report an event occurrence
3. *Information.* An FE used to provide data that is used by an application

An FE is produced based on the requirements of the applications and according to the SCAI models. Each FE is described using the template shown in Figure 5.24.

The protocol is written based on the FE descriptions. Information that is specifically translated into a protocol is contained in a parameter list. Each parameter in the list has three aspects. Let's identify and discuss them.

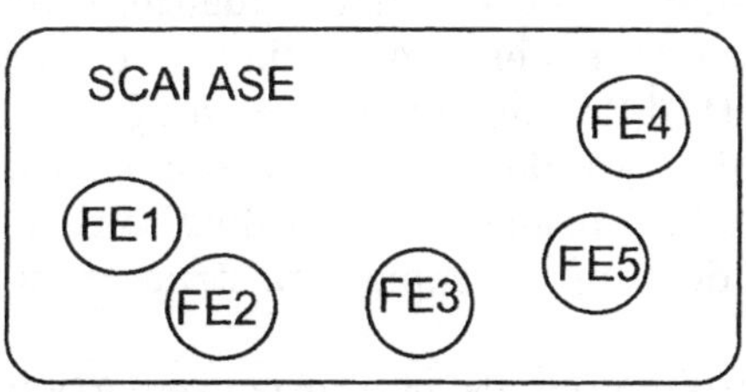

Figure 5.23 SCAI ASE FEs.

Direction: The direction of the FE—switch to computer *OR* computer to switch

Description: Explains the function of the FE

Operational Rules: Rules and procedures governing operation of the FE

Call Model State Transitions (for event FEs): Indicate call model transitions during which the FE is issued; also, Party Condition transitions are indicated.

OR

Allowable Call States (for all other FEs): Indicate the states which allow issuance of the FE.

Parameter List: Lists all parameters of the FE

	Name	*Type*	*Req*	*Ind*	*Rsp*	*Cnf*
Request:						
Response:						
Success:						
Error:						

Parameter Definitions: Definitions of all parameters

Transition Causes (for event FEs): List of state transitions causes

OR

Error Definitions (for all other FEs): Definitions of all errors associated with the FE handling

Figure 5.24 FE template.

1. Each parameter has a name and a type. The parameter's type is later defined using ASN.1 notation. The type is important because some parameters that have different names can be easily identified as actually representing the same notions if their type is the same. For example, there are parameters with different names—*received_Call, established_Call, delivered_Call*—that are defined by the same type: Call_ID. This technique allows naming parameters according to their functional role in a specific FE, but defining the same data structure to represent them in a protocol.

2. Each parameter is placed in either a request or a response portion of the template. The request/response notion is taken from ROSE (which is actually used to transport SCAI FEs to a remote peer). Fig-

ure 5.25 shows the ROSE operation model as it applies to SCAI. The model indicates that every package of information (message) is sent to a remote system as a request. A reply with either a success or failure may be returned* depending on the type of operational class of the SCAI service.

3. Each parameter is mapped onto one of the OSI primitives: Req, Ind, Rsp, Cnf. This mapping helps in identifying the usage of each parameter in OSI terms.

All FEs are grouped into *functional groups* according to their functions. The functional groups are a descriptive, not a protocol, tool. They are used to classify the types of services that the SCAI standard supports.† The functional groups are:

- Monitor/Filter
- Basic Two-Party Call
- Basic Two-Party Call Progress
- Agent Events
- Conference
- Transfer
- Hold/Retrieve
- Features
- Routing
- Call Diversion
- Other

5.5.2 SACF rules

In Chap. 4 we introduced the standard application layer structure, including the concept of SACF. The rules contained in the SCAI Con-

* A question may arise as to why the term *Response* is used in the SCAI standard instead of the term *Reply* from ROSE. This was done to stress that the FE template shows SCAI services. The SCAI services are mapped onto ROSE services. *Success* and *Error* from the template are also SCAI terms.

† Initially, the concept of functional units was considered, but there was no agreement on it. A functional unit, in OSI terms, is a subset of an ASE functionality. In SCAI, a functional unit would contain a number of FEs, just like a functional group does. The difference is that at the association time functional units can be negotiated; that is, two systems may decide which functional units to use. A functional group, on the other hand, cannot be negotiated. The concept of functional units was not adopted in SCAI for several reasons. One of the reasons is lack of sufficient experience in the industry with functional units and their usefulness.

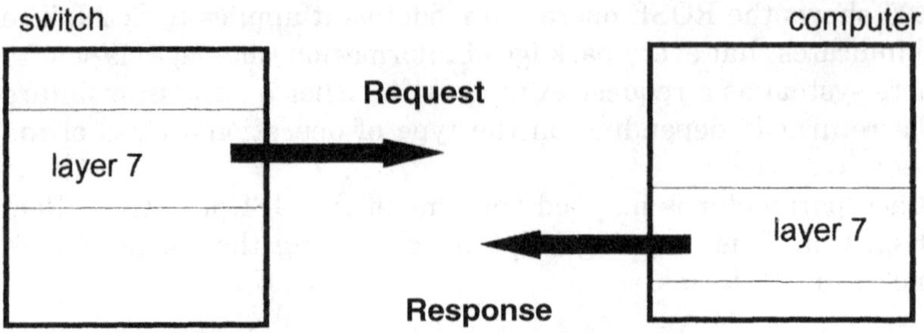

Figure 5.25 ROSE model.

trol Function are sufficient for the successful operation of the SCAI ASEs. These rules are:

1. ACSE must be the first ASE to be accessed within an ASO in order to set up an association.
   MEANING: When an application process wants to establish communication with another application process, it has to establish an association. The mechanism to do this is an ACSE. Without first establishing an association, no SCAI operations can be performed.
2. ACSE must be the last ASE to be accessed within an ASO in order to release the association.
   MEANING: When SCAI operations are complete for the purposes of a specific application process or a fatal error forces the end of SCAI operations, the ACSE is responsible for releasing the previously established association.
3. SACF coordinates all requests within the SAO.
   MEANING: All requests for service issued to any ASE in the SAO are coordinated by the SACF. For example, if the application process needs to request a specific service from the SCAI ASE, this request is passed to the SACF and then to the SCAI ASE.
4. The SCAI ASE accesses ROSE.
   MEANING: The SCAI ASE uses ROSE for exchanging information with a remote peer. All SCAI FEs are shipped across the link to a destination system as part of ROSE PDUs.
5. The SCAI ASE does not access presentation services.
   MEANING: ROSE is the SCAI's mechanism for interfacing with lower layers of the OSI stack. The SCAI standard does not allow the SCAI ASEs to access the lower layers, starting with the presentation layer, without the services of ROSE.
6. ACSE and ROSE do not interact.
   MEANING: Since ACSE and ROSE are created to perform separate tasks, there is no need for their interaction.

The SACF rules are part of the SCAI standard and they govern general operations of the SAO containing SCAI ASEs. The particular implementation of SACF, including the portion of it that interfaces with the user's application process, is quite open to an implementor.

5.5.3 SCAI ASE FEs

We will discuss SCAI FEs as part of functional groups, but first I would like to present one of the FEs and explain how each entry of the FE template is used (see Fig. 5.26).

Direction: Computer→Switch

Description:
The client requests that an existing call be disconnected and related resources be freed. FE class = Request.

Operational Rules:
This service maps to a ROSE class 2 operation. If the request is invalid, no action is taken with respect to devices in the call. When the call is cleared no further event reports shall be reported for the call. The call identifier is to be released.

Either the call identifier or the device requesting the call to be cleared shall be specified (as part of party identifier). If the call identifier of the call to be cleared is not specified, and only one call exists at the requesting/clearing device, then that call is cleared. If more than one call exists at the requesting/clearing device and the call to be cleared cannot be unambiguously identified, an error is returned and no call is cleared.

To insure that a specific call is cleared, the call identifier of that call should be provided. If the call identifier is not supplied, there is a possibility that an unintended call may be cleared.

Allowable Call States: PENDING, ORIGINATED, DELIVERED, ESTABLISHED, FAILED.

Parameter List:

	Name	*Type*	*Req*	*Ind*	*Rsp*	*Cnf*
Request:	party_to_Clear	Party_ID	M	M(=)		
	private_Data	SCAI_Private_Data	U	C(=)		
Response:						
Success:	private_Data	SCAI_Private_Data			U	C(=)
Error:	error	Error_Groups			M	M(=)

Parameter Definitions:

party_to_Clear	identifies the call to be cleared or the device to clear the call
private_Data	identifies custom data specific to an implementation or application
error	identifies the response error

Error Definitions:

Error Groups:

1. General_Errors
2. Clear_Call_Errors
3. Private_Errors

Figure 5.26 Clear_Call FE description.

Clear_Call_Errors:	
unknown clearing device	the device cannot be determined
unknown call to clear	the call cannot be determined
invalid clearing device	the device cannot clear the call
invalid call to be cleared	the call cannot be cleared at the device
no call present	no call is present at the device
multiple calls present	the request cannot be completed because it is not clear which call is to be cleared

Figure 5.26 *(Continued)*

The FE is Clear_Call. It is sent by a computer to a switch. The description identifies the FE's functionality. This FE is used to request a disconnect of a specific call. The class of the FE is Request. The fact that it is a request may be clear from the description of this particular FE, but in other cases this classification may help a designer to understand a function faster.

Operational rules are descriptions of the behavior in specific instances. Their number varies from FE to FE because the circumstances surrounding activation of each FE are very specific. The more errors can occur in the situation when a particular FE is used, the more operational rules are defined. Each FE's operational rules begin with its mapping to a class of ROSE operation. The idea of mapping to ROSE operations is based on the desire to indicate what kind of response is expected when an FE is issued. If an FE is mapped to a class 2 operation, it means that a success or failure response is expected to be returned. If the mapping is to a class 5 operation, there is no response expected. For the Clear_Call FE, either a SCAI-level success or failure is returned. Most of the rest of the operational rules are devoted to the handling of error conditions.

Allowable call states are those call states within which an FE can be invoked. For Clear_Call, these are PENDING, ORIGINATED, DELIVERED, ESTABLISHED, FAILED. An application can issue a Clear_Call FE and request that a particular call be cleared only when the call is in one of these states.

The Clear_Call FE contains two parameters:

1. party_to_Clear. This identifies the call to be cleared or the device which is involved with a call to be cleared. The type (the data structure) of this parameter is Party_ID, which is the same as the type of the party-related parameters in other FEs. This parameter is mandatory; the value of the indication primitive is the same as the value of the request primitive.
2. private_Data. This is a field that allows specific implementations of the SCAI standard to pass nonstandard information between sys-

tems. It is defined to permit creation of nonstandard solutions for some of the users' needs.* The type of this parameter is SCAI_Private_Data. The parameter is optional in the request primitive and is present in the indication primitive if it is present in the request primitive. The value of the parameter in the indication primitive is equal to that in the request primitive.

If the success is returned, there is one parameter present: private_Data. It is optional, so the success may be returned with no parameters at all.† If private_Data is present in the response primitive, it is also present in the confirm primitive and has the same value as in the response primitive.

If the error is returned, a single parameter is mandatory in the response primitive and in the confirm primitive with the value equal to that in the response primitive. This parameter is *error* of the type Error_Groups. Since Error_Groups is a type that is referenced in most FEs, it is worth discussing in detail. This type consists of three error groups:

1. General_Errors
2. Specific_FE_Errors
3. Private_Errors

General errors are those that apply to many FEs and are not usually related to specific FEs' parameters. These are:

Privilege violation: e.g., a particular request is not allowed

Resources not available or a particular resource is not available

Other error: e.g., all errors not specifically defined in the standard

Private errors are those that were generated by information carried by the private data parameters. Just as private data are different for each implementation, so are private errors.

The errors specific to each FE are defined in the FE. In the case of Clear_Call FE, several errors are defined and the definitions are self-explanatory.

---

* All switch and computer manufacturers attempt to build features into their equipment that would differentiate them from their competitors. These features are usually proprietary and guarded by their creators, although that is not always the case. It is customary to define so-called "escape fields" in standards to allow a small degree of nonstandard functionality. In SCAI, the private data parameter can be used to pass application information between switches and computers in support of this need.

† This is an important point. The fact that a success response has been returned, even with no parameters, is sufficient for the application to understand that the operation is successful. Parameters present in the response add information.

A word about mandatory and optional parameters: All parameters that are critical to the understanding of the SCAI environment in a given situation are mandatory. For example, in the Clear_Call FE, the Party ID parameter is absolutely necessary. Without identifying either a call or a device it is impossible to clear a call. This parameter is mandatory; its absence would force an application to completely ignore the request. Private data, on the other hand, is an optional parameter. There are two reasons to have optional parameters. First, a parameter is optional when it is not always used, as in the case with the private data parameter in the Clear_Call FE. Second, a parameter is optional if it cannot be guaranteed that the information necessary for this parameter is always available. For example, in the Call_Originated FE, the Cause parameter is optional because systems can exist that do not possess an ability to identify the precise cause of the transition (see the appropriate section on Call_Originated FE). In addition, this parameter is not critical to the operation of an application. So, if the computer application expects the Cause parameter from the switch, but the switch is unable to provide it, the application has to take some special actions; however, this fact should not mean that the application cannot continue with the call.

The following sections contain descriptions of all SCAI FEs, but only to the degree necessary to either present a function or add clarification to the standard's text. These sections are not intended as a copy of the SCAI standard's text.

Monitor/Filter functional group. Monitoring is a key concept in SCAI. It provides the ability for applications to receive the information necessary for their operation. This functional group provides two types of functions:

1. *Monitoring*—setting up monitors for specific devices
2. *Filtering*—selecting specific reports to be issued for monitored devices

Monitoring is the process of identifying SCAI objects for which event reports are to be issued and reporting the events occurring at these objects. When a monitor is set for a specific device, all activities at that device become visible to an application that set the monitor. This visibility allows an application to make decisions based on the specific application's requirements. In the SCAI standard, monitoring is defined for a device object. Other objects (call, agent, party) cannot be monitored at this time. This is done because calls, parties, and agents are all attributes of the device object and device monitoring allows visibility of activities related to all those objects.*

---

* Parties, agents, and calls (contained in parties) are attributes of the individual device object. Since a group device consists of members, which are individual devices, we can say that by monitoring members of a group device, the same visibility effect can be achieved.

What does it mean to set up a monitor? It is a process of identifying a device to be monitored and assigning a monitor reference value that can be used to correlate monitors among themselves and with other reference values (call IDs, etc.). A single monitor reference value (monitor ID) can be assigned to more than one monitored device if an application requires this. Two methods for setting up monitors are allowed in the SCAI standard:

1. *Explicit monitor.* Explicit monitors are established dynamically. After an application association is established, specific FEs are used to indicate the devices to be monitored. Events are reported right after the request for monitoring is accepted by a switch.
2. *Implicit monitor.* Implicit monitors are established prior to the establishment of an application association via a subscription mechanism. Subscription is an ability to predefine (provision) which are the devices to be monitored and how to reference them. Events are reported after the establishment of an association.

Filtering is a function of selecting specific events to be reported for the monitored devices. A large number of event reports is already defined in the SCAI standard and more event reports can be added later. Not all applications need to see all events as they happen. A specific application may be interested in some events, but not in others, and this mix can change in time and from application to application. Filtering permits an application to receive only those event reports that are of interest. When a monitor request is issued (or when monitored devices are set up by subscription), a filter is also issued. Events are reported according to the list provided in the filter.

Filter is a dynamic function. The types of events that an application requires "seeing" can be easily changed by issuing a filter for a specific device. When a new filter request is processed, events are reported accordingly.

In Fig. 5.27, the monitored telephone device is observed by the application. When there is a call at this device, all transitions of the call through the states of the call model are visible to the application. In general, a monitored device does not have to be a telephone. It can be any device that behaves by the rules established for the device object.

Five FEs are defined in the Monitor/Filter functional group:

1. Initiate_Monitor
2. Cancel_Monitor
3. Change_Monitor_Filter
4. Query_Monitor
5. Monitor_Report.

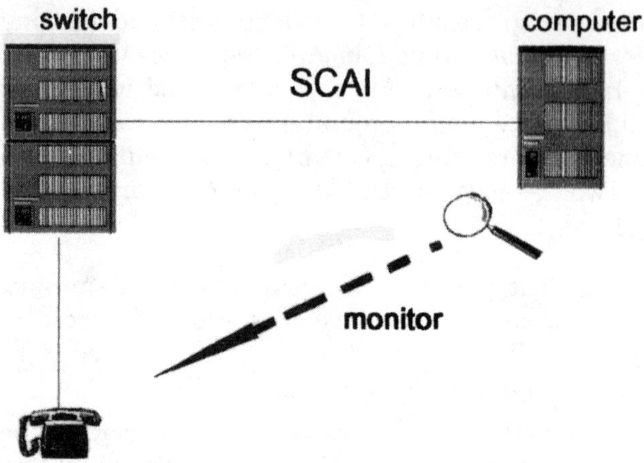

Figure 5.27 Monitoring a device.

Since all FEs in this group are not call-related (they do not affect calls), there is no reference in their descriptions of the allowable call states or call state transitions.

Initiate_Monitor. The Initiate_Monitor FE sets up monitoring of a particular device. When the request is issued, three parameters can be included:

- *Device:* A mandatory parameter indicating which device is to be monitored.
- *Requested filter:* An optional parameter indicating the reports to be sent to the application; if this parameter is included, a specific list of event reports is sent to a switch; if this parameter is excluded, a switch uses a preset default filter.
- *Private data:* Optional field for private data.

If the monitoring request is accepted, the success is returned with two parameters:

- *Monitor ID:* A reference assigned by the switch to a monitored device; management of the monitor IDs is accomplished by the switch based on its internal implementation and on the needs of the application.
- *Private data.**

---

* Private data is included in the request and applicable successful response portions of each FE; therefore, there is no need to explain its use for every FE. It is always present to provide an implementation field for information exchange, and it is always optional.

If an error is discovered and it is critical to the understanding of the request, monitoring is not activated and a specific error is returned in the response. If the problem is the inability to issue specific reports because they are not supported by the application context, the default filter can be returned; the default filter is an optional parameter in the data structure ERRORS (see ASN.1 for this FE).

So far we have discussed setting monitors for individual devices. If the number of devices is large, a large number of Initiate_Monitors has to be issued, which is not very efficient. However, if individual devices are combined into device groups, a single request for monitoring can be issued identifying the group as a monitored device. In this case, not only activity at the group would be seen, but an activity at each member of the group. The effect is that one monitor request covers a number of individual devices. Of course, all members of the group will generate the same kind of report. If different reports are expected from different members, an application either has to issue individual monitor requests or a single monitor request for the entire group with subsequent requests to change filters on specific devices (in which case it is an additional burden on the application to correlate members of the same group generating different reports with that group ID).

Cancel_Monitor. As its name implies, this FE cancels previously set monitors. No reports are issued on the indicated device after the cancel request is processed. When the cancel request is issued, two parameters can be included:

- *Monitor ID:* A mandatory parameter that identifies a monitor to be canceled. When a monitor ID was assigned, it was based on the knowledge of the device ID for which a monitor request has previously been issued. Clearing a monitor ID clears a monitor at the specific individual or group device.
- *Private data.*

Change_Monitor_Filter. This FE carries a request to change a filter on a previously established monitor. It can contain up to three parameters in its request portion:

- *Monitor ID:* A mandatory parameter that identifies the monitor to be affected.
- *New filter:* A mandatory parameter that identifies a new list of events to be reported. If these events are not supported by the application context, a default filter may be returned as part of the error parameter (see ASN.1 for this FE).
- *Private data.*

Query_Monitor. This FE is used to request the status of the established monitors, or, in other words, to learn of the correlation of monitor IDs to device IDs and to the filters used for the monitors. It can be issued any time an application needs to update its information, but it was put in primarily as the mechanism for relating the presubscribed monitor/filter data to the starting application. This FE can include two parameters in the request portion:

- *Monitor ID:* An optional parameter identifying a monitor for which a report is requested. If it is not present, reports on all established monitors are sent.
- *Private data.*

This FE is linked with the Monitor_Report FE as a parent. When Query_Monitor is received, a switch either responds with a linked Monitor_Report, or, if an error was discovered, returns Query_Monitor_Error with a mandatory error identification parameter.

Monitor_Report. This is either the response FE to the Query_Monitor or an FE sent by the switch based on some application criteria. It provides the information of interest to the application. The request portion of this FE can contain four parameters:

- *Monitor ID:* A mandatory parameter identifying a monitor.
- *Device:* A mandatory parameter identifying a device related to the monitor ID.
- *Filter:* A mandatory parameter identifying a list of event reports to be issued for this device.
- *Private data.*

If an error is discovered in the report, a mandatory error parameter is returned. In the case of an error, an application decides whether to reissue a request for information or not.

Basic two-party call functional group. The FEs in this group provide services necessary for the basic calls. A basic call in SCAI is a call without associated supplementary services. A basic call can be set up or released. Manipulations of a call, such as transfer or conference, are considered to be supplementary to the basic calls. Although the functions necessary for basic calls seem to be simpler than, let's say, transfer, the majority of calls are basic and require reliable and efficient support.

Four FEs are in this group:

1. Make_Call
2. Answer_Call
3. Clear_Call
4. Predictive_Make_Call

Make_Call. This FE contains a request to a switch to establish a two-party call between two given devices. A computer indicates the originating and terminating devices to be involved in the call, but it is up to a switch how to proceed with the call attempt: which route to select, priority treatment, and so on. This is a typical third-party call control FE; it triggers a function that will affect a device on whose behalf the FE is issued.

Make_Call can be invoked in the NULL and PENDING call states (originating half of the call model). This is an FE that is mapped to a ROSE class 2 operation (a response is always returned, either with success or failure). The request portion of this FE contains up to four parameters:

- *Calling device:* A mandatory parameter indicating a device on whose behalf the call is being placed.
- *Called device:* A mandatory parameter indicating a device that is the destination for the attempted call.
- *Call type:* An optional parameter identifying the type of call: voice or data; if missing, the default is voice.
- *Private data.*

If the call request is accepted by the switch, a successful response is returned with the following two parameters:

- *New Call:* A mandatory parameter carrying a call reference value assigned by the switch, call ID; the call ID is used in all call-related FEs to reference this particular call. Management of call IDs is not a SCAI function and is performed by a switch based on its implementation.
- *Private data.*

Answer_Call. This FE is issued when there is a need to answer a call on behalf of a called device. It is usually deployed in situations when a called device is a hands-free device (headset). This FE can be issued in the ARRIVED and RECEIVED call states (terminating half of the call).

The request portion of the FE contains two parameters:

- *Party to answer:* A mandatory parameter indicating the party ID associated with the destination device; the party ID contains either a call ID of the call to be answered or a device ID of a called device (or both).
- *Private data.*

Clear_Call. This FE requests that an existing call be disconnected. If disconnect is performed, the related resources (devices, trunks, etc.) are cleared. This FE can be issued in all states of the originating call model (except NULL) and in the ESTABLISHED and FAILED states of the terminating call model.

The request portion of this FE contains two parameters:

- *Party to clear:* A mandatory parameter indicating the party ID of the call to be cleared; a party ID can contain either a call ID of the call to be cleared or a device ID of the affected device (or both).
- *Private data.*

Predictive_Make_Call. This FE performs a function very similar to the Make_Call FE: it requests a call setup to a specified device. The difference is that the Predictive_Make_Call requests a connection to the calling device (the device on whose behalf the call is attempting to be established) only after the called device has answered. This FE is used to set up calls between a calling group device (such as an ACD group) and a called device.

When a called device has answered, a switch places the call at the group device for the distribution to the agent according to the algorithm used. This can happen when an outgoing call is in the DELIVERED or ESTABLISHED call state. A parameter in the FE specifies the state.

Predictive_Make_Call can only be invoked in the NULL state.* The request portion of the FE contains up to five parameters:

- *Calling device:* A mandatory parameter indicating a device on whose behalf the call is being originated.
- *Called device:* A mandatory parameter indicating a destination device.
- *Call type:* An optional parameter indicating a voice or data call.

* The functionality of predictive make calls is related to group devices that distribute originating calls. Hence, the invocation is possible in the NULL state only.

- *Distribution:* An optional parameter indicating the call state within which a call is to be distributed to an originating individual device (e.g., an agent). If the parameter is absent, the state is DELIVERED.
- *Private data.*

In the case of the acceptance of the request, a success response is returned with two parameters:

- *New call:* A mandatory parameter identifying a call ID.
- *Private data.*

Basic two party call progress functional group. This functional group contains FEs that report the occurrence of the call-related events. Only two-party basic call events are reported by the FEs in this group. The FEs that report events related to call activity other than basic two-party calls are contained in different functional groups.

One of the FEs in this functional group is issued every time there is a transition of a call from state to state. When an event report FE is received by a computer, an application makes a processing decision based on the information in the FE. One informational item that the description of this kind of FE contains is the cause for transition. This information is necessary for an application's decision making.

Eight FEs, referenced as event reports, are defined in the basic two-party call progress functional group:

1. Service_Initiated
2. Call_Originated
3. Call_Delivered
4. Call_Arrived
5. Call_Received
6. Call_Established
7. Call_Cleared
8. Call_Failed

All FEs in this group are mapped to a ROSE class 5 operation; that is, no responses to requests are issued.

Service_Initiated. This FE reports a service request event. In many instances the service request signifies a device going off-hook in the attempt to originate a call (on a standard phone, a user would hear a dial tone). Since this is an originating call attempt, only the originating half of the call model is affected: the call transitions from NULL to PENDING. The originating party condition transitions from NULL to ACTIVE, while the terminating party condition does not change because at this point there is no terminating party associated with the call.

This FE contains five parameters:

- *Monitor ID:* A mandatory parameter identifying a reference value of the monitor established for the device involved with this call.
- *Initiated call:* A mandatory parameter indicating a call ID just assigned by the switch to this call.
- *Initiating device:* An optional parameter identifying a device that initiated the call.
- *Cause:* An optional parameter indicating a specific cause for the event (i.e., a service request).
- *Private data.*

Call_Originated. This FE reports an event that is the completion of the dialed digits collection and the initiation of the call to another (terminating) device. Two call state transitions are possible:

1. From PENDING to ORIGINATED, with the originating party condition staying unchanged. This is a typical way of transitioning to ORIGINATED.
2. From NULL to ORIGINATED, with the originating party condition transitioning from NULL to ACTIVE. This transition occurs, for example, when the originating device is always attempting a call to the same number and, therefore, there is no need to dial any digits; the switch is programmed to recognize that this particular device always calls the same number. Another possible example is the knowledge by the switch that when this particular device is attempting a call, the switch itself selects a termination device, meaning that no digit collection is required.

This FE contains six parameters:

- *Monitor ID:* A mandatory parameter identifying a reference value for the monitor established for the originating device involved with this call.
- *Originated call:* A mandatory parameter indicating a call ID of the originating call.
- *Calling device:* An optional parameter indicating an initiating device.

- *Called device:* An optional parameter indicating a terminating device.
- *Cause:* An optional parameter identifying the cause of transition.
- *Private data.*

Call_Delivered. This FE indicates that a terminating device is alerting (a telephone which is the destination for the call is ringing). This FE reports an event that occurred on the terminating side of the call as a transition in the originating call model. There is a comparable transition and an event report related to the terminating call model (see Call_Received FE).

A call transitions from the ORIGINATED to the DELIVERED state. The initiating party condition stays unchanged, while the called (terminating) party condition transitions from NULL to ACTIVE.

This FE contains seven parameters:

- *Monitor ID:* A mandatory parameter identifying a reference value of a monitor established for the device initiating this call.
- *Delivered call:* A mandatory parameter indicating an ID of the call.
- *Delivered device:* An optional parameter indicating a device at which the call is being terminated.
- *Calling device:* An optional parameter indicating an initiating device.
- *Called device:* An optional parameter indicating a device which was called; it may be different from the delivered device if the call was forwarded or for some other reason.
- *Cause:* An optional parameter indicating the cause of the transition.
- *Private data.*

Call_Arrived. This FE reports the arrival of a terminating call. This is the beginning of a call termination attempt by a switch. This state cannot be reflected in any actions observed at a standard phone by a user in the way a dial tone is. The state is necessary for SCAI because it provides an application with call progress details and information regarding the calling party. Two call state transitions are possible:

1. From ROUTING to ARRIVED, when a terminating call is first suspended due to the need to perform a routing function (ROUTING state) followed by the return to the normal call processing.*

---

* By "normal" I mean call progress that does not include transitions that bypass some states of the call model.

2. From NULL to ARRIVED, when no routing is required. No change in the party conditions is taking place.

This FE contains seven parameters:

- *Monitor ID:* A mandatory parameter identifying a reference value for the monitor established for the intended terminating device.
- *Arrived call:* A mandatory parameter indicating the ID of the call.
- *Destination device:* A mandatory parameter indicating the device intended to receive the call.
- *Calling device:* An optional parameter indicating the device that is calling.
- *Called device:* An optional parameter indicating a device that was originally called; it can be a different device from the one at which the call is intended to terminate (as a result of call forwarding, etc.).
- *Cause:* An optional parameter identifying the cause of the transition.
- *Private data.*

Call_Received. This FE reports that the preparation for call delivery to a terminating device is completed and the terminating device (phone) is alerting (ringing). Two call state transitions are possible:

1. From ROUTING to RECEIVED, when there is no need to validate whether the selected device is appropriate.
2. From ARRIVED to RECEIVED, when the call goes through the normal transitions.

In either case, the terminating party transitions from the NULL to the ACTIVE condition; there is no change in the originating party condition.

This FE contains seven parameters:

- *Monitor ID:* A mandatory parameter identifying a reference value for the monitor established for the intended terminating device.
- *Received call:* A mandatory parameter indicating the ID of the call.
- *Received device:* A mandatory parameter indicating the terminating device.
- *Calling device:* An optional parameter indicating the calling device.
- *Called device:* An optional parameter indicating the called device; it can be different from the received device due to actions like call forwarding.

- *Cause:* An optional parameter indicating the cause of the transition.
- *Private data.*

Call_Established. This FE reports that a terminating device has answered the call and the conversation is now possible. This indicates the end of a call setup procedure. Four call state transitions are possible:

1. From ORIGINATED to ESTABLISHED, when there is no need to alert the terminating device. This is an originating call model transition. The terminating party transitions from the NULL to the ACTIVE condition, while the originating party condition does not change.
2. From DELIVERED to ESTABLISHED, when the call is going through the normal transition. This is an originating call model transition with no changes in both parties' conditions.
3. From ARRIVED to ESTABLISHED, when there is no need to alert the terminating device. This is a terminating call model transition. The terminating party transitions from the NULL to the ACTIVE condition, while the originating party condition does not change.
4. From RECEIVED to ESTABLISHED, when the call is going through the normal transition. This is a terminating call model transition with no changes in both parties' conditions.

This FE contains seven parameters:

- *Monitor ID:* A mandatory parameter identifying a reference value of the monitor established for the originating or terminating device, depending on either originating or terminating call in progress.
- *Established call:* A mandatory parameter indicating the ID of the call.
- *Answering device:* An optional parameter indicating the answering device.
- *Calling device:* An optional parameter indicating the calling device.
- *Called device:* An optional parameter indicating the called device. It can be different from the answering device due to actions like call forwarding.
- *Cause:* An optional parameter identifying the cause of the transition.
- *Private data.*

Call_Established FE has an interesting property: it is used to report transitions in both originating and terminating call models. Of course, two event reports could have been defined, such as Call_Established_Originated and Call_Established_Terminated, but this was found to be unnecessary. The result is that the FE contains a description of transitions in originating and terminating call models, and, more importantly, the parameter designations could not be clear without additional operational rules. The problem is that, depending on the half of the model that is affected, one or another parameter must be present (mandatory). The operational rules state that, for an originating call, the calling device is known and must be provided,* and for a terminating call the answering device is known and must be provided. These operational rules are covered in the protocol.

Call_Cleared. This FE reports the end of the call. Usually this indicates that all resources associated with this call were released, but, since the FE is also used to drop one of the parties from a conference call, the operational rules specifically state that, in the case of a conference, if the dropped device is not a conference controller, the call may continue to exist (see *conference functional group* description for more information).

This FE indicates a call transition from any state to NULL. Accordingly, both parties transition to the NULL condition. This FE contains five parameters:

- *Monitor ID:* A mandatory parameter identifying a reference value of the monitor established for the affected device.
- *Cleared call:* A mandatory parameter indicating the ID of the cleared call.
- *Clearing device:* An optional parameter indicating the device that initiated call clearing.
- *Cause:* An optional parameter indicating the cause for this transition.
- *Private data.*

Call_Failed. This FE reports that a call cannot be completed due to a discovered error. An example of a typical failure is when the called party is busy (therefore, the call cannot be completed) and a busy tone is heard by the calling party. There are five possible transitions to the FAILED state, from PENDING, ORIGINATED, DELIVERED, ARRIVED, and RECEIVED.

---

* It is customary for the word *shall* in the standards to indicate that a function or a parameter must be supplied.

This FE contains six parameters:

- *Monitor ID:* A mandatory parameter identifying the reference value of the monitor established for affected device.
- *Failed call:* A mandatory parameter indicating the ID of the failed call.
- *Calling device:* An optional parameter indicating the calling device.
- *Called device:* An optional parameter indicating the called device.
- *Cause:* An optional parameter identifying the cause of the transition.
- *Private data.*

Agent events functional group. This is the agent-related functional group that contains the FEs' reporting agent events. As discussed previously, the current need for agent information is derived from the requirement to collect statistics regarding agents' operations, performance of individual agents, and so on. In addition, we can imagine an application that is built around the knowledge of agents' names—for example, when Ms. Jones is signed on device #55, do not route regular calls to her because she is acting as a supervisor. In general, an application can use agent-related information for whatever it finds necessary. The reports of the events associated with agents' status change are created for these purposes.

There are six FEs in this functional group:

1. Agent_Logged_On
2. Agent_Logged_Off
3. Agent_Ready
4. Agent_Not_Ready
5. Agent_Working_Ready
6. Agent_Working_Not_Ready

All FEs in this functional group are mapped to a ROSE class 5 operation, which means that no response is issued to any requests. All FEs contain exactly the same parameters and, therefore, the parameters are described only with the first FE. Since these FEs are reporting events occurring with non-call-related objects (agents), state transitions are not indicated in the FE descriptions.

Agent_Logged_On. This FE reports that the agent log-on operation has been invoked. An agent is signing on the device. The FE contains five parameters (the same as in other FEs of this functional group):

- *Monitor ID:* A mandatory parameter identifying the reference value of the monitor established for the affected device.
- *Agent device:* A mandatory parameter indicating the device that an agent is logging on to (for other FEs, it indicates a device affected by an action).
- *Agent ID:* An optional parameter identifying the agent.
- *Agent group:* An optional parameter indicating the group an agent is a member of.
- *Private data.*

Agent_Logged_Off. This FE indicates that the agent log-off operation has been invoked. An agent is signing off the device.

Agent_Ready. This FE reports that an agent is ready to accept calls.

Agent_Not_Ready. This FE reports that the agent is not available to receive calls.

Agent_Working_Ready. This FE reports that the working ready operation has been invoked. The agent is still occupied with processing the previous call-related information, but can receive new calls.

Agent_Working_Not_Ready. This FE reports that the agent working not ready operation has been invoked. The agent is still processing the previous call-related information and cannot receive new calls.

Feature functional group. This functional group is supposed to contain functions that are concerned with observing and manipulating device features. A *device feature* is any feature associated with the device itself. It is envisioned that a computer sends a request to a switch to treat a feature in/at the device in a specific way. An example of a feature that can be set at a device is call forwarding. When call forwarding is set at a device, calls for this device are forwarded to another device. A computer sends a request to set or clear a feature for a particular device and a switch performs the necessary actions to either accept or reject the request. However, the present standard contains provisions only for requests associated with agent states transitions. For example, if an agent in a READY state needs to transition to a NOT READY state, he or she issues a request on a computer terminal and a computer passes this request to a switch; a switch changes the state of the agent.

Two FEs are contained in this functional group:

1. Set_Feature
2. Query_Feature

These FEs are issued without regard to call states because at the present time they affect only objects not directly involved with calls—the agent object. Hence, allowable states are not indicated in the FEs' descriptions.

Set_Feature. This FE requests a switch to set or clear a feature on a device. The request portion of this FE contains three parameters:

- *Device:* A mandatory parameter identifying the device on whose behalf a feature is being set or cleared.
- *Set feature information:* A mandatory parameter indicating which feature is affected and information related to this feature.
- *Private data.*

Information supplied in the request portion of this FE identifies the feature (operation) to be performed on a specific device and may contain agent-related information, depending on the request and the application requirement for that information.

Query_Feature. This FE requests the current status of a feature on a device. Because currently only agent-oriented features are covered in the SCAI standard, the operational rules for this FE address only agent-related procedures. When more features are added, the operational rules will be changed to reflect the needs of other features.

The request portion of this FE contains three parameters:

- *Device:* A mandatory parameter identifying the device at which a feature is queried.
- *Feature:* A mandatory parameter indicating the feature that is queried.
- *Private data.*

The success response portion of this FE contains two parameters:

- *Query feature information:* A mandatory parameter presenting feature-related information.
- *Private data.*

Hold-Retrieve functional group. This functional group contains FEs that perform the actions of putting calls on hold and retrieving the calls from hold, and FEs reporting these occurrences. A call is placed on hold when it is necessary to accept or originate another call from the same device. A call is retrieved when the user at the device desires to return to a call placed on hold. The requests to hold and retrieve calls are performed by a computer on behalf of the devices indicated in the FEs.

Four FEs are included in this functional group:

1. Hold_Call
2. Call_Held
3. Retrieve_Call
4. Call_Retrieved

Hold_Call. This FE is issued on behalf of a device to interrupt an existing call at that device and place a party at this device on hold. Hold is a condition that means that a party associated with the device requesting a hold is not participating in any information exchange. The call associated with the party that was put on hold is not disconnected and can be retrieved later. This FE can be issued in three states: ORIGINATED, DELIVERED, and ESTABLISHED. It contains two parameters in the request portion:

- *Party to hold:* A mandatory parameter identifying a party to be held; either a device ID or a call ID must be specified.
- *Private data.*

Since this FE maps to a ROSE class 2 operation, there is a response expected to the issued request. A successful response can contain one optional parameter—private data.

Call_Held. This FE reports that a call has been placed on hold. No state transition occurs, but a condition of the party involved with this call at the device that requested the hold changes from ACTIVE to HOLD. The party can be either originating or terminating.

The request portion of the FE contains four parameters:

- *Monitor ID:* A mandatory parameter identifying a reference value of the monitor established for the affected device.
- *Held call:* A mandatory parameter indicating the ID of the affected call.
- *Holding device:* A mandatory parameter indicating the device at which the hold function has been executed.
- *Private data.*

Retrieve_Call. This FE is issued on behalf of a device to re-establish a call which is being held. The request can be issued in three states: ORIGINATED, DELIVERED, and ESTABLISHED. Two parameters are contained in the request portion of the FE:

- *Party to retrieve:* A mandatory parameter indicating the party to be retrieved.
- *Private data.*

Call_Retrieved. This FE reports that a call has been retrieved (re-established). No state transition occurs, but the party condition changes from HOLD to ACTIVE. Four parameters are contained in the request portion of this FE:

- *Monitor ID:* A mandatory parameter identifying a reference value of the monitor established for the affected device.
- *Retrieved call:* A mandatory parameter indicating the ID of the retrieved call.
- *Retrieving device:* A mandatory parameter indicating the device at which a call has been retrieved.
- *Private data.*

Transfer functional group. This functional group contains FEs that perform functions concerned with call transfers. Generally speaking, transfer is a function that results in a call being transferred from one device to another. It is widely used in applications that SCAI supports, such as a customer service center. Calls are transferred to an agent who is more qualified to deal with a customer, or to a supervisor, or to a different department. Two types of transfer are supported by this functional group: *single-step* and *consultation.* In a single-step transfer, an existing call is disconnected and a new call attempt is generated to a different destination. In a consultation transfer, two existing calls are disconnected and transferred into an established new call.

This functional group contains three FEs:

1. Single_Step_Transfer
2. Consultation_Transfer
3. Call_Transferred

Single_Step_Transfer. This FE is issued on behalf of a device to transfer a call to another device that is not currently participating in the call. The result of this action is that an existing call is sent to another party, and the party on whose behalf the request was issued is disconnected from this call. Whether the new party (the party to which the call is transferred) can answer the call may not be known at the time of the transfer request.

This FE can be issued only in the ESTABLISHED state of the call to be transferred. The request portion of this FE contains four parameters:

- *Transferring device:* A mandatory parameter indicating the device from which a call is to be transferred.
- *Destination device:* A mandatory parameter indicating the device that is the destination for a transferred call.

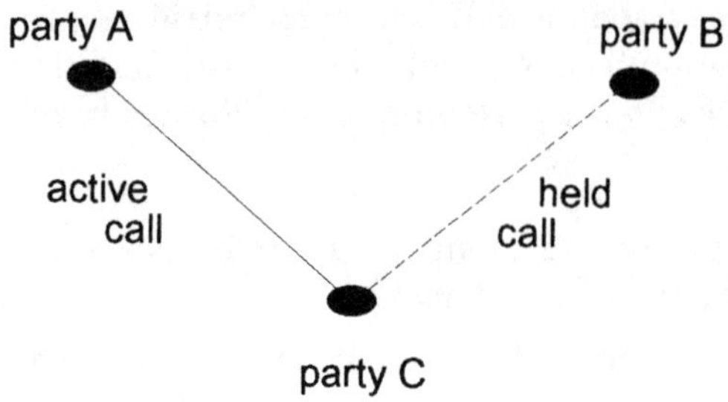

Figure 5.28 Consultation transfer (part 1).

- *Established call:* An optional parameter identifying the ID of a call to be transferred.
- *Private data.*

The successful response portion of the FE contains two parameters:

- *New call:* An optional parameter indicating the ID of the "new" call, the call that is attempting to connect to the destination device.
- *Private data.*

Consultation_Transfer. This FE is issued on behalf of a device to establish a new call between two parties involved in two other calls, one of which is active and the other on hold. The party on behalf of which the call is being transferred is currently a party to both calls, but, after consultation transfer is executed, this party terminates its participation in either call.

In Fig. 5.28, party C was involved in a call with party B and put it on hold. Party C is currently communicating with party A. After the execution of the consultation transfer request on behalf of party C, parties A and B are communicating and party C is disconnected from both calls (see Fig. 5.29).*

---

* Strictly speaking, using the SCAI object definitions would force us to address each combination of a device and a call as a separate party. If agent Jones answers a call from customer Smith while located at device #20, the party ID associating the device with a call is, let's say, XX. The same agent Jones at the same device calling a supervisor is now generating a new device/call combination and, therefore, a new party ID, let's say, YY. In our example from Figure 5.28 we should address party C as party C1 involved in the active call, and as party C2 involved in the held call. Following this very formal method would create a very complex description of the discussed procedures. To avoid this, I am addressing both parties, C1 and C2, as party C, implying that there is a user (agent) hidden behind this party ID (name). I hope it is easier for the reader to understand functions like transfer if the description, although not strictly formal, is closer to the real-life examples that we all experience. Hence, I use phrases like "a party is involved in call 1 and call 2. . . ."

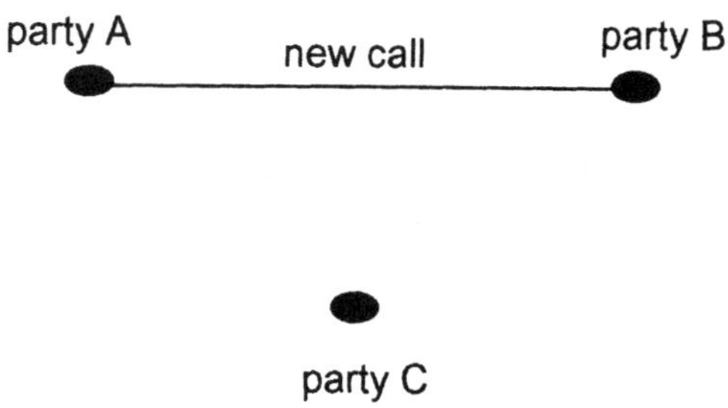

Figure 5.29 Consultation transfer (part 2).

This FE can be issued only in the ESTABLISHED state, and it contains four parameters in its request portion:

- *Transferring device:* A mandatory parameter indicating the device on whose behalf the calls are to be transferred.
- *Call 1:* An optional parameter identifying the first call (either of the two calls can be the first or the second) at the transferring device; the other party in this call (party A) will be a party to the new call established after the transfer.
- *Call 2:* An optional parameter identifying the second call at the transferring device; the other party in this call (party B) will be a party to the new call established after the transfer.
- *Private data.*

The successful response portion of this FE contains two parameters:

- *New call:* An optional parameter indicating the ID of a new call (between parties A and B).
- *Private data.*

Call_Transferred. This FE reports that a call has been transferred from one device to another. This event is reported for the device that requested the transfer or on whose behalf the request was issued.

Three cases of transferred calls are possible, depending on the origination of the transfer request. The list of parameters in all three cases is the same (it is introduced after the discussion of differences among the cases).

Case 1. When a transfer request is issued through the computer and the request is for the consultation transfer, the Call_Transferred FE reports the following state transitions of the calls involved in the transfer:

- Both existing calls transition from ESTABLISHED to NULL
- A newly established call transitions from NULL (it did not exist just a moment ago!) to ESTABLISHED.

Let's now observe what is happening with the parties involved in this procedure. Three parties are involved: A, B, and C (see Fig. 5.30).

Party C was involved in two calls, which are indicated in Fig. 5.30 as active and held. After the transfer, these calls do not exist anymore (that is why they are shown in parentheses), which means that party C transitions to the NULL condition for both calls (these previous calls are indicated in the FE as parameters *call 1* and *call 2*). Party C is the transferring party.

Party A was involved in one previous call, *call 1—active*. Since, after the transfer, call 1 does not exist any more, party A, associated with this call, transitions from any party condition it happened to be in to NULL. Since party A now participates in a new call, its condition for this new call is the same as that of the party associated with the previous call at this device. If the condition of the party associated with call 1 was active, the condition of the party associated with a new call is also active. Party A did not have any association with another call (call 2) and, therefore, its condition as it relates to call 2 did not change. Party A is the transferred party (although it can also be a transferred-to party; we need to assign these labels to discuss the example).

Party B was involved in one previous call, *call 2—held*. Since, after the transfer, call 2 does not exist any more, party B, associated with this call, transitions from any condition it happened to be in to NULL. Since party B participates in a new call, its condition for this call is the same as the condition of the party associated with the previous call at this device. Party B did not participate in call 1 and, therefore, its condition as it relates to call 1 did not change. Party B is a transferred-to party.

Case 2. When a transfer request is issued through the computer and the request is for the single-step transfer, the Call_Transferred FE reports the following state transitions:

- A current call that is being transferred transitions from ESTABLISHED to NULL

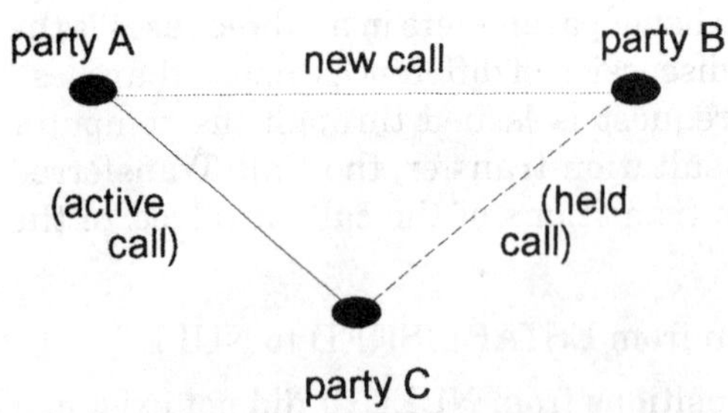

Figure 5.30 Three parties in a transfer.

- A new call transitions from NULL to one of the following states: ORIGINATED, ARRIVED, or ROUTING. It transitions to the ORIGINATING state if the call is an originating call. It transitions to either ARRIVED or ROUTING, depending on the requirements of the application, if the call is a terminating call.

The parties associated with the existing call transition from any to the NULL condition (the call is disconnected). The condition of the party that is supposed to participate in a new call is unchanged (it is not related to a new call yet). The condition of the party at the device involved in existing and new calls is unchanged.

Case 3. This case is referred to in the standard as a manually invoked transfer. "Manually" means that a function is invoked at the telephone set. The result is that a computer application is not involved in requesting that a certain function be executed and does not know anything about it. The ability to invoke functions from the telephone terminal is available to users, but the application has to be informed about the action. Case 3 reports the occurrence of the transfer invoked from the telephone set when one of the calls was already established and the other call was either established or in the process of being established (remember, in case 1 both calls were established). Let's follow the state transitions for all involved calls.

Call 1 (the call that has to be established) transitions from ESTABLISHED to NULL. Call 2 also transitions to NULL from either ESTABLISHED (if the call was established), or ORIGINATED (if it is an originating call), or DELIVERED (if it is a terminating call).

A new call transitions from NULL to one of the following states, depending on the state of call 2 (e.g., if call 2 is in the ESTABLISHED state, the new call transitions to the same state): ORIGINATED, DELIVERED (originating calls), ROUTING, ARRIVED, RECEIVED (terminating calls), ESTABLISHED (any calls).

All party conditions transitions are the same as in case 1.

Call_Transferred FE contains seven parameters:

- *Monitor ID:* A mandatory parameter identifying a reference value of a monitor established for a transferring device.
- *Transferring device:* A mandatory parameter indicating the device from which calls are being transferred.
- *Destination device:* An optional parameter indicating the destination for the transferred call (new call).
- *Call 1:* An optional parameter indicating the ID of one of two calls involved in the transfer, in the case of a single-step transfer, this parameter contains the call ID of the transferred call.

- *Call 2:* An optional parameter indicating the ID of the other call; it is never used in the case of a single-step transfer.
- *New call:* An optional parameter indicating the ID of the new call.
- *Private data.*

Conference functional group. This functional group contains FEs that perform the functions necessary for creating and manipulating a conference. The first phase of the SCAI standard defines conferencing functions that allow only up to three parties to participate in a conference. Although this is clearly a deficiency of the first phase of the standard, the three-party conference is used in many applications that SCAI supports. In making the decision between delaying the standard and omitting more general conferencing capabilities, the latter won. A more general capability is planned for the second phase.

There are two conferencing terms that should be defined here:

- *Conference controller.* A device that controls the conference; usually it is the device that sets up a conference, but it does not have to be so; this varies depending on a particular application.
- *Conference participant.* A noncontrolling device participating in a conference.

The process of creating a conference in SCAI is the process of merging two existing calls—one active and one on hold (Fig. 5.31).

Party C places a call with party B on hold and establishes a call with party A. After the conference request is executed, all three parties can communicate: they are the participants in a conference call (Fig. 5.32).

Four FEs are contained in this functional group:

1. Conference_Existing_Calls
2. Drop_Conference_Party
3. Calls_Conferenced
4. Conference_Party_Dropped

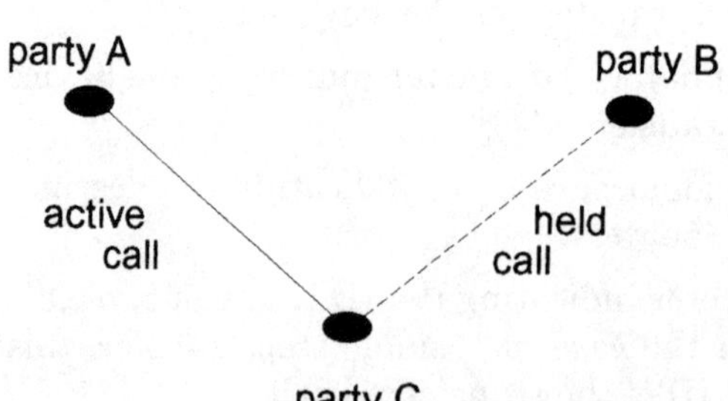

Figure 5.31 Conferencing (part 1).

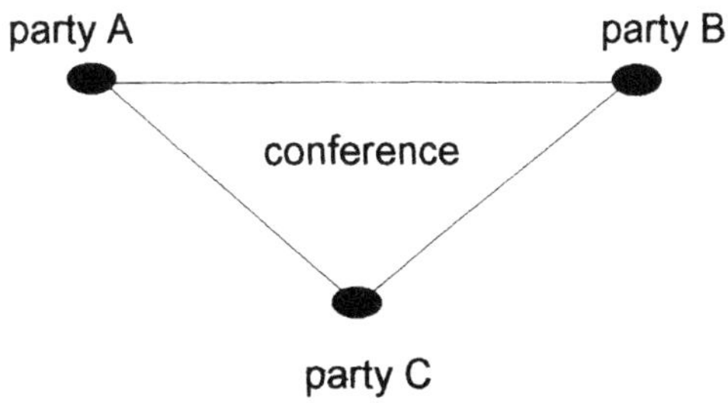

Figure 5.32 Conferencing (part 2).

Three FEs from other functional groups are also used in conferencing: Hold_Call, Retrieve_Call, and Clear_Call. The first two FEs are used to place one party on hold in the process of creating a conference and then later retrieve it. The last FE can be used to disconnect one of the parties from the conference. Also, related event reports are used: Call_Held, Call_Retrieved, and Call_Cleared.

Conference_Existing_Calls. This FE is issued to join two existing calls into one conference. This request can be issued only in the ESTABLISHED state. The request portion of the FE contains four parameters:

- *Conferencing device:* A mandatory parameter indicating a device common to both calls to be conferenced (party C).
- *Call 1:* An optional parameter identifying the first of two calls to be joined (it is not specified in SCAI which of the calls is the first or the second).
- *Call 2:* An optional parameter identifying the second of the two calls to be joined.
- *Private data.*

The successful response portion of the FE contains five parameters:

- *Conference:* A mandatory parameter identifying the ID of the conference; SCAI does not specify whether this ID is the same as one of the conferenced calls or if it is a new value.
- *Controller:* An optional parameter identifying a conference controller.
- *Party 2:* An optional parameter identifying the party from call 1.
- *Party 3:* An optional parameter identifying the party from call 2.
- *Private data.*

Drop_Conference_Party. This FE is issued on behalf of a conference controller to drop a participant from the conference. Any party can be dropped using this FE, including a controller itself. It can be issued only in the ESTABLISHED state.

Three parameters are contained in the request portion of the FE:

- *Conference:* A mandatory parameter identifying the ID of a conference.
- *Party to drop:* A mandatory parameter identifying the party to be dropped.
- *Private data.*

Calls_Conferenced. This FE reports that two existing calls have been joined into a single three-party conference. The event is reported only for the device corresponding to the conference controller. The conference condition becomes active. Among the three parties now participating in the conference, the party associated with the conference controller becomes active, while the two other parties' conditions stay unchanged.

The request portion of this FE contains nine parameters:

- *Monitor ID:* A mandatory parameter identifying the reference value of a monitor established for a device associated with a conference controller.
- *Conferencing device:* A mandatory parameter indicating the device at which two calls have been conferenced.
- *Call 1:* A mandatory parameter indicating one of the two calls that have been joined.
- *Call 2:* A mandatory parameter indicating the other call that has been joined.
- *Conference:* A mandatory parameter indicating the ID of the conference.
- *Controller:* An optional parameter identifying a party that is associated with the conference controller.
- *Party 2:* An optional parameter identifying a conferencing party from call 1.
- *Party 3:* An optional parameter identifying a conferencing party from call 2;
- *Private data.*

Conference_Party_Dropped. This FE reports that a party was dropped from the conference. It is reported only for the conference controller. If a conference participant other than the controller is dropped using a Drop_Conference_Party FE, the event is reported by a Clear_Call FE. The reason for this is the desire to clear a call ID associated with the dropped party. If party B from Fig. 5.32 is dropped, the original call between parties B and C is also cleared.

The transitions of the conference and involved parties depend on whether the dropped party is a conference controller. If it is a conference controller, then the following transitions take place:

The condition of the conference changes from ACTIVE to FLOAT; if, after the conference controller drops out, there is only one party left in the conference or no party is left in the conference, the conference condition transitions to NULL (the conference is cleared).

The condition of the party which is associated with the conference controller changes to NULL.

The condition of other parties is unchanged.

If the dropped party is not a conference controller, then the following transitions take place:

The condition of the conference does not change.

The condition of the conference controller does not change.

The condition of the dropped party changes to NULL.

This FE contains four parameters in the request portion:

- *Monitor ID:* A mandatory parameter identifying the reference value of a monitor established for the affected device.
- *Conference:* A mandatory parameter indicating the conference ID.
- *Dropped party:* A mandatory parameter indicating the party that has been dropped.
- *Private data.*

Routing functional group. This functional group contains the FEs necessary to obtain routing information for a call—i.e., finding an appropriate destination. Usually, a switch that handles terminating calls has sufficient information to decide what is the destination for each call. If call distributors are involved, it is their goal to find an appropriate destination for a call. However, there are applications that may want to perform a selection of a destination on their own, based on the criteria known to them. In these instances, regular call processing in a switch would be suspended for a call while a computer is searching for the destination address. If information from the computer is not received by the switch within a predetermined time, call processing is reactivated and the call is completed using regular switch procedures.

This functional group contains three FEs:

1. Route_Request
2. Route_Selected
3. Routing_Toggle

Route_Request. This FE requests a computer to identify the destination for the call. It is issued only in the NULL state of the terminating call model. While it is not an event report FE, it does report the transition from the NULL to the ROUTING state. Call processing for the identified call is suspended (so the ROUTING state can be called a wait state) for a specific length of time, and the response is expected in the Route_Selected FE. There is no success response portion of the Route_Request FE defined; the response is sent only when an error is discovered.

This FE contains five parameters in the request portion:

- *Call to route:* A mandatory parameter identifying the call for which the destination is sought.
- *Destination device:* A mandatory parameter indicating the destination device where the call is to be routed in case the computer does not respond with a new destination.
- *Calling device:* An optional parameter identifying the device that originated the call.
- *Called device:* An optional parameter identifying the device that was originally called; it may not be the same as the destination device.
- *Private data.*

Route_Selected. This FE is issued by a computer to respond to the switch's request for the destination selection. It contains three parameters:

- *Call to route:* A mandatory parameter identifying the call for which the route request was issued.
- *Alternate destination device:* An optional parameter identifying a new destination for the call; if this parameter is omitted, the switch routes the call to the original destination device. Note that if a computer does not assign a new destination and does not issue a Route_Selected FE, the switch waits for this information until a timeout occurs and, therefore, wastes time.
- *Private data.*

Routing_Toggle. Sending a Route_Request FE to receive routing information from a computer can be part of the static arrangement for an application (preprogrammed) or can be requested dynamically. Routing_Toggle FE is issued to invoke this function dynamically. When this FE is received by the switch, the calls for a specified destination begin transiting through the ROUTING state. This FE is used to dynamically activate or deactivate the routing function. It is a non-call-related FE; the calls are affected by it only after a switch receives it and begins sending routing requests to a computer.

The FE contains three parameters:

- *Device:* A mandatory parameter identifying the device for which routing is enabled; it can be an individual or a group device.
- *Routing activate:* A mandatory parameter indicating activation or deactivation of the routing function.
- *Private data.*

Call Diversion functional group. This functional group is supposed to contain FEs performing call diversion functions. Unfortunately, only one FE has been defined. Additional FEs in this group are planned for future releases of the SCAI standard. In general, call diversion functions are those that divert calls from one device to another based on some condition. For example, call forwarding is such a function and it can be invoked based on a busy condition of a device, or based on the time of day, or the like.

Call_Diverted. This FE reports that a call has been diverted from one device to another. Diversion can occur on the terminating half of the call before the call reaches the ESTABLISHED state. When a call is diverted, it transitions from ARRIVED, RECEIVED, or FAILED to the NULL state. The FE contains six parameters:

- *Monitor ID:* A mandatory parameter identifying a reference value of the monitor established for the affected device.
- *Diverted call:* A mandatory parameter identifying the diverted call.
- *Diverting device:* A mandatory parameter indicating the device from which the call has been diverted.
- *New destination:* An optional parameter indicating a new call destination.
- *Diversion cause:* An optional parameter indicating the cause of diversion.
- *Private data.*

Since at the present time there are no functions defined in the SCAI standard to request call forwarding or other diversion-type services, this FE can only be used with equipment that does include one or more of those functions. In that case, only reporting of the event is performed in a standard fashion, not the function itself.

Other functional group. This functional group was created to contain FEs that do not belong in any other category. At this time, only one FE is defined: Network_Reached.

Network_Reached. This FE is issued when a call has left the SCAI domain (the SCAI operational space). It implies that the situation has changed and there may not be information forthcoming regarding the further progress of this call. The FE can be used in situations when a signaling system in the portion of the network that a call is entering cannot supply the expected terminating end information. The application would have to make adjustments based on some preset procedures.

Network_Reached does not indicate any call state transitions. It contains five parameters:

- *Monitor ID:* A mandatory parameter identifying the established monitor.*
- *Existing call:* A mandatory parameter identifying the call.
- *Calling device:* An optional parameter indicating the originating device.
- *Called device:* An optional parameter indicating the destination device.
- *Private data.*

In the works: The *Conferencing* function will be changed. More than three parties will be allowed to have a conference, in effect creating conferencing of *n* parties. A *Call Forwarding* function will be added.

5.5.4 SCAI security functions

Security is a critical element of the SCAI standard. It provides necessary protection against intruders on what is, in most cases, a very sensitive interface—an interface between a network and the outside world. With the proliferation of intelligent devices and the development of relatively simple methods of writing programs for those devices, the potential for abuse of information access interfaces is growing. This is why network and information service providers pay a great deal of attention to the security aspects of their interfaces and deploy equipment that can support a secure working environment.

---

* Note that if the call is leaving the switch on a trunk, the monitor has to be set at the trunk (or a trunk group). Since at present time "trunk" is not an explicit SCAI object, an application dealing with trunks would have to find a solution for this problem. One way of dealing with this is to treat trunks as any other devices, in which case care has to be exercised in setting up an event-reporting mechanism.

Two types of security breaches can occur with SCAI:

1. Network-related
2. Customer-related

Network-related breaches are those that attempt to compromise network integrity. An intruder may want to achieve one of the following results:

- Gain access to the network's operation systems (management systems) to acquire sensitive information about the network
- Use network resources without authorization (getting a service without paying for it)
- Sabotage network components through actions such as flooding the network with service requests or messages

and others.

Customer-related breaches are those that attempt to compromise a customer's integrity. An intruder may want to achieve one of the following results:

- Acquire information on sensitive traffic patterns and clients
- Destroy or contaminate data in customers' computers
- Gain access to sensitive market-related information

and others.

I do not attempt to fully present the types of breaches that are possible, but just to outline the seriousness of the security issue when anybody deploys an interface extending from a network out. Some of the mentioned potential problems are handled by existing switch and computer security functions such as passwords. Other potential problems require specific coverage. SCAI provides support for the following security requirements:

1. *Peer-Entity Authentication.* The ability to supply the identity of the peer application process.
2. *Data Origin Authentication.* The ability to supply the identity of the source of the message.
3. *Connection Integrity.* The ability to detect any insertion, deletion, or replay of any data within an entire message sequence.
4. *Connection Confidentiality.* The ability to maintain the confidentiality of user data.

5. *Nonrepudiation with Proof of Origin.* The ability to prove that a message was sent and thus reject an attempt by the sender to deny sending the message.

Peer entity authentication is supported by ACSE via the exchange of appropriate parameters at the association time. The functions necessary to support the rest of the security requirements are performed by the SCAI Security ASE. This ASE contains a single FE—SS_Transfer—which performs an encapsulation of ROSE APDUs into secure envelops.

The Security ASE is activated when a ROSE APDU with SCAI information contained in it is submitted for transmission to a peer system, or when a ROSE APDU is received from a peer system. The Security ASE either encapsulates the PDU in a secure envelop or disassembles security information from a received PDU (see Fig. 5.33).

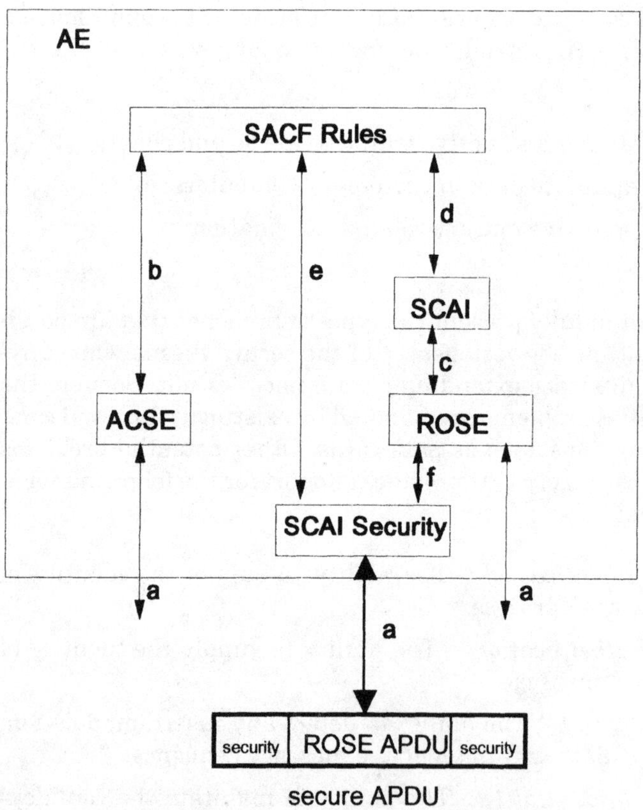

Figure 5.33 Secure PDU.

SS_Transfer. The role of SS_Transfer is to provide security functions for SCAI. The parameters of this FE are used to provide information necessary for security. These parameters are:

- ROSE APDU: This is the PDU itself that is to be secured; it must be present.
- *Sender ID:* An optional parameter providing an ID of the sender; this parameter supports two requirements: data origin authentication and nonrepudiation.
- *Confidentiality algorithm:* An optional parameter identifying the algorithm used by the sender for encryption; this parameter supports the requirement for connection confidentiality.
- *Sequence number:* An optional parameter that provides a sequence integrity service; it supports the requirement for connection integrity.
- *Time stamp:* An optional parameter providing an indication of the time the message was sent; this parameter supports two requirements: connection integrity and connection confidentiality.
- *Integrity check:* A mandatory parameter that contains a digital signature and the algorithm identifier used for the signature; this parameter supports three requirements: data origin authentication, connection integrity, and nonrepudiation.

The combination of parameters used in a particular arrangement depends on the security needs. Some requirements can be supported by more than one parameter, and the choice of these parameters is up to the implementor.

Let's consider an example of using SCAI security functions. First, when an application association is being established, the ACSE authentication function confirms that the peer application entity is the one with whom the communication was meant to be established.

When SCAI APDU is created (let's say a Make_Call is issued), it is mapped onto ROSE Invoke APDU and then passed to the SCAI Security ASE for security processing. Let's imagine that all optional parameters of SS_Transfer are used.

When this secure APDU is received by the peer system's SCAI Security ASE, it processes security information. If no problems are found, the ROSE APDU is passed to ROSE for further processing; otherwise, the application process (through the SACF) is informed of problems. The other SS_Transfer parameters are examined as follows:

The sender ID is examined to compare it with the entity ID with which the communication was intended. It is also recorded as proof of sending the message for nonrepudiation purposes.

The confidentiality algorithm is used to decode the encrypted ROSE APDU; the encryption was used to protect the message from an unauthorized reading of its contents.

The sequence number is used to check if the message arrived in the appropriate sequence.

The time stamp is used to check the message against other messages to find out if it has been substituted (with a correct sequence number) for another message sent at a different time.

The integrity check is used to detect any alteration of the message through the decoding of the digital signature.

5.6 Application Contexts

Application contexts identify the functions to be used during communications over a particular association. Each standardized application context has an object ID assigned to it by the appropriate standards committee. At the association time, these IDs are exchanged by systems planning to communicate with each other. An object ID points to a predetermined set of functions. After two systems agree on a specific application context, only the functions defined in that application context can be used.

Two application contexts are defined in the SCAI standard:

1. Application Context One contains:
   SCAI ASE
   ROSE
   ACSE
   SACF Rules
2. Application Context Two contains:
   SCAI ASE
   SCAI Security ASE
   ROSE
   ACSE
   SACF Rules

The difference between the two application contexts is the SCAI Security ASE. The availability of two application contexts allows implementors to choose the application context depending on whether or not they require security functionality.

The object ID of SCAI Application Context One has the following value*:

---

* In Sec. 4.2.1, we discussed that the object ID value consists of the sequence of all values corresponding to the nodes in the path for an object. Accordingly, we should expect a value next to each node in the path. However, the nodes starting three branches—*ccitt, iso, joint-iso-ccitt*—are allowed to be referenced by name only, e.g., iso.

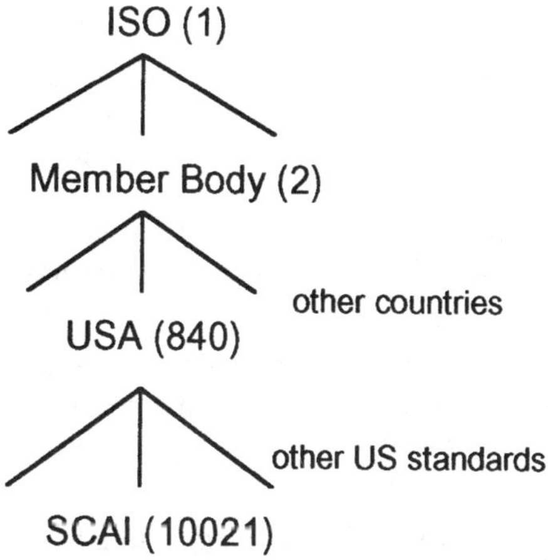

Figure 5.34 Assignment of the SCAI object ID.

```
{iso member-body(2) uSA (840) sCAI(10021) sCAI-context(8)
version-one(0)}*
```

The object ID of SCAI Application Context Two has the following value:

```
{iso member-body(2) uSA (840) sCAI(10021) sCAI-security-
context(9) version-one(0)}
```

The SCAI field in the object ID name, whose value is "10021," distinguishes the SCAI standard from other U.S. standards. The next field after "sCAI" distinguishes object IDs within the SCAI standard. The value of this field in Application Context One is "8," and in Application Context Two is "9." Other values in these object ID fields are assigned for ISO (=1), member body (=2), USA (=840). The path for the SCAI object ID is shown in Fig. 5.34.

Assignment of the object ID for the SCAI standard required an allocation by ANSI, which is the U.S. registration authority, of the number which is now used for the assignment of object IDs for all SCAI-related objects (e.g., application contexts, ASEs, etc.).

---

* It is clear to a reader who possesses the text of the SCAI standard that the object IDs for all objects in the standard are different from the ones presented here. The object IDs in the SCAI standard are incorrect. How this error sneaked into the standard is difficult to say; perhaps it was the same sort of mistake that created some other minor inconsistencies and errors. At the time of writing this book, the contribution to T1S1 is also being written to correct this error.

5.7 SCAI Protocol

The SCAI protocol is specified using the ASN.1 notation. The encoding rules for the abstract syntax used in the SCAI standard are Basic Encoding Rules (BERs) with a few restrictions. These restrictions are:

1. The encoding of the bit string value and an octet string value must be primitive (although the BER standard allows the constructed form also).
2. For the definite form of the length, the long variant of the encoding can be used only if the number of the octets is more than 127 (although the BER standard allows the use of the long variant for any number of octets).
3. In the long variant of the length encoding, it is not permissible to use more octets than the minimum necessary for the correct representation (although the BER standard does not have this restriction).

The SCAI abstract syntax (SCAI protocol) is defined in six modules. These modules are:

- SCAI abstract syntax module one
- SCAI abstract syntax module two
- SCAI ASE module
- SCAI Security ASE module
- SCAI Application Context module one
- SCAI Application Context module two

The decision on the number of modules was made primarily based on the desire to enhance readability of the text. Whether the goal was achieved or not is up to the users of the standard to decide, but, at the time, this structure looked logical.

All six protocol modules have object IDs assigned to them so that any module can be referenced by any other module. In addition, there are four other object IDs assigned to the SCAI objects:

- SCAI ASE
- SCAI Security ASE
- SCAI Application Context One
- SCAI Application Context Two

These object IDs are required to uniquely identify the aforementioned objects for global use (we already discussed the last two objects in the previous chapter). When SAOs are created to support an application, or when an association is being established, these are the object IDs that are referenced.

5.7.1 SCAI Abstract Syntax Module One

The SCAI Abstract-syntax-module-one contains definitions for data types corresponding to all SCAI operations and errors. Its object ID is:

```
{iso member-body(2) uSA(840) sCAI(10021) sCAI-abstract-
syntax-module-one(0)version-one(0)}
```

SCAI operations are functions, FEs, which we discussed in Sec. 5.5 from the functional point of view. An implementor of the SCAI standard actually implements the protocol defined in this and other SCAI ASN.1 modules.

Although all operations and errors are defined in this single module, each operation is placed in a separate chapter for convenience. Also, the operations are grouped into functional groups to correspond to their descriptions as FEs. Obviously, when this module is prepared for parsing by an implementor, the chapter headings for each functional group and each operation must be excluded or coded as comments. Otherwise, the ASN.1 parser will generate errors.

This module imports OPERATION and ERROR macros from ROSE. Both macros are defined in ROSE and used in the definition of each SCAI operation. This module exports all data types defined for the SCAI operations so they can be referenced by other modules. It is also worth noting that some common data types are actually collected together (in a single chapter) for convenience. These are types for Party ID, Device ID, and so forth, all used as components of other structured data types.

Let's consider the abstract syntax of the Clear_Call operation as an example (see Fig. 5.35).

The function Clear_Call is defined as type OPERATION, which itself is defined in ROSE and consists of three components: ARGUMENT, RESULT, and ERRORS (these components correspond to the portions of the Clear_Call FE: Request, Success Response, and Error Response).

ARGUMENT is defined by ClearCallArgument, which is a sequence of two components: partyToClear and privateData (and these components correspond to two parameters in the request portion of the Clear_Call FE). The component party ToClear is of the type PartyID, which is defined in a different place in the standard—chapter 10.3.1.12 of T1.626. The component privateData is of the type SCAIPrivateData, which is also defined in chapter 10.3.1.12 of the SCAI standard. This parameter is indicated as being OPTIONAL.

```
ClearCall::=     OPERATION
                 ARGUMENT   ClearCallArgument
                 RESULT     ClearCallResult
                 ERRORS     {ClearCallError, GeneralError,
                                              PrivateError}

ClearCallArgument::=SEQUENCE
    {
         partyToClear           PartyID,
         privateData            SCAIPrivateData        OPTIONAL}

ClearCallResult::=CHOICE
    {
         noData                 NULL,
         privateData            SCAIPrivateData}

ClearCallError::=ERROR

         PARAMETER              ClearCallErrorParameter

ClearCallErrorParameter::=SEQUENCE
    {
         errorNumber            ClearCallErrorNumber,
         errorInformation       Errorinformation       OPTIONAL}

ClearCallErrorNumber::=ENUMERATED
    {
         unknownClearingDevice              (0),
         unknownCallToClear                 (1),
         invalidClearingDevice              (2),
         invalidCallToBeCleared             (3),
         noCallPresent                      (4),
         multipleCallsPresent               (5)}
```

Figure 5.35 Clear Call ASN.1.

RESULT is defined by ClearCallResult, which is a choice of either noData (defined as NULL—no data present in the response) or privateData.

ERRORS is defined by the list consisting of three types. GeneralError and PrivateError types are defined separately in chapter 10.3.1.13 of the SCAI standard. ClearCallError is defined as type ERROR, which in turn is defined in ROSE. ERROR contains a single component—PARAMETER, which is defined by ClearCallErrorParameter as a sequence of ClearCallErrorNumber and Errorinformation. ClearCallErrorNumber is of the ENUMERATED type and contains a list of errors with assigned values. This list corresponds to the list of FE-specific errors in the Clear_Call FE description. Errorinformation is optional and defined in chapter 10.3.1.13, together with the rest of the general errors.

Abstract syntax for each operation is defined based on the corresponding FE description. There are a couple of instances when parameters of FEs are not exactly the same as corresponding components of data types in the abstract syntax describing these FEs. But this is more a case of an unintended difference. In any respect, any parameter of an FE can be always mapped to a data type component, even if it looks different at first glance.

5.7.2 SCAI Abstract Syntax Module Two

SCAI Abstract-syntax-module-two defines the values for the types of all SCAI operations and errors. Its object ID is:

```
{iso member-body(2) uSA(840) sCAI(10021) sCAI-abstract-
syntax-module-two(5)version-one(0)}
```

All operation types and error types are imported from the SCAI abstract-syntax-module-one; all assigned operation and error values are exported for use by other modules. This module also imports the SS-Transfer type from SCAI Security-ase-module and exports the assigned value for this type.

The assigned values are used to distinguish all operations and all errors. For example, when operation value "7" is received, an application understands it to be a request for a "single-step transfer," and operation value "13" indicates the "call established" event report.

5.7.3 SCAI ASE Module

SCAI Ase-module defines the operations of the SCAI ASE in terms of consumer and supplier. Consumer and supplier, in general, may or may not invoke the same operations; it is possible to define an ASE in which only consumer or supplier invokes operations. In the case of the SCAI ASE, both consumer and supplier invoke operations, but not the same operations. While the assignment of roles (consumer, supplier) is arbitrary, in the SCAI standard a computer is a supplier (it issues—supplies—commands) and a switch is a consumer (it receives—consumes—commands and issues reports). The object ID of this module is:

```
{iso member-body(2) uSA(840) sCAI(10021) sCAI-ase-module(1)
version(0)}
```

This module imports the values of all SCAI operations from the SCAI abstract-module-two and the definition of APPLICATION-SERVICE-ELEMENT (ASE) macro from ROSE. It exports the value of the SCAI ASE, which is:

```
{iso member-body(2) uSA(840) sCAI(10021) sCAI-ase(6) version-
one(0)}
```

This module corresponds to the prose definition of the SCAI ASE as a collection of functions (FEs). The operations that the consumer invokes correspond to the FEs that have the switch→computer direction. The operations that the supplier invokes correspond to the FEs that have the computer→switch direction.

5.7.4 SCAI Security ASE Module

The SCAI Security-ase-module defines types and values associated with security services. Its object ID is:

```
{iso member-body(2) uSA(840) sCAI(10021) sCAI-security-ase-
module(2)version-one(0)}
```

This module exports the security operation (SS-Transfer) and the value of the SCAI Security ASE, which is:

```
{iso member-body(2) uSA(840) sCAI(10021) sCAI-security-ase(7)
version-one(0)}
```

The module imports the following:

- ASE macro definition from ROSE.
- The value of the security operation type from the SCAI abstract-syntax-module-two.
- Type DistinguishedName from AuthenticationFramework (it is defined in the X.501 Recommendation). This type is used to reference IDs of senders (applications, users, etc.) registered in global databases (see X.500-series of ITU-T Recommendations).
- Types AlgorithmIdentifier, SIGNED, and ENCRYPTED from AuthenticationFramework* (it is defined in X.509 Recommendation). SIGNED and ENCRYPTED are macros that provide notations for secure operations.

* The SCAI standard contains a typo in the AuthenticationFramework object ID. The node denoted "de(5)" is actually the same node as in the InformationFramework—"ds(5)".

SS-Transfer is defined as a SIGNED SEQUENCE type. The application of SS-Transfer to a PDU can produce either an encrypted or unencrypted version of the data stream. It is specified by the choice of either sCAI-operation or confidential-sCAI-operation (which is encrypted). The SIGNED macro contains operations for generating a digital signature.

To finish the presentation of this module, I just want to mention the type that we have not discussed previously—UTCTime. It is a built-in type intended to specify universal time required for international applications.

5.7.5 SCAI Application Context Module One

The SCAI Application-context-module-one defines the Application Context One that we discussed in the previous chapter. The object ID value of this application context is exported.

5.7.6 SCAI Application Context Module Two

The SCAI Application-context-module-two defines the Application Context Two that we discussed in the previous section. The object ID value of this application context is exported.

5.8 Conformance to the SCAI Standard

We have talked about various SCAI issues: from architecture to functions, from modeling to protocol, and so on. We need now to discuss what would constitute conformance with the SCAI standard. What are the criteria by which we can state whether an implementation is compliant with the standard or not.

Usually, a standard contains two types of text: normative and informative. It is obligatory to comply with what is identified as normative text. Informative text is usually provided as a description or guidance and does not contain directives that require compliance.

In the SCAI standard, it is stated that certain sections of the document require compliance in order to claim that an implementation conforms to the standard. These sections are:

- 6.1 and 6.2—use of ROSE and ACSE
- 8—SCAI functional elements
- 9.2—SCAI security functionality
- 10—SCAI protocol

Sections 6.1 and 6.2 of the standard specify how ROSE and ACSE are to be used. Section 8 specifies all SCAI functions. These functions are repeated in section 10 in the form of a protocol that is to be implemented and, therefore, a question can be asked regarding the value of requiring conformance not only with section 10, but also with section 8. Section 8 provides critical information that is not captured in the protocol: operational rules. The operational rules are as important for compatibility between systems as the protocol elements, and must be followed for conformance.

Although modeling and architectural sections are not mentioned in the list of sections for conformance, they do impact the protocol sections. For example, each FE is specified in terms of the SCAI objects; the parameters are defined based on the knowledge about these objects. Hence, conformance with the protocol also implies conformance with the object model.

Section 9.2 and a part of section 10 specify the security functions and protocol. Conformance to these sections is required if the SCAI security functionality is to be used, which brings us to another question of conformance: whether it is required that each function (FE) defined in the standard be implemented to claim conformance. The answer to this question is "No." It is not necessary to implement each function. Some of them, such as event reports, can be turned off even if they are implemented. Others, such as make_call, can never be used even if they are implemented. The point of the standard for information exchange is not to demand the implementation of each defined function, but to make operations of the functions that are implemented compatible in different equipment. It is an important point that the SCAI standard, and other standards of this type, are not standard protocols for transfer of data. Instead, this is a protocol for the exchange of information that is used by applications. Because of this fact, it is desirable to allow implementation only of those functions that are required in a specific arrangement. Another aspect of the same fact is the difficulty of performing conformance testing.

Conformance testing is performed to compare the operation of a system with a standard. In the case of a transport protocol such as X.25, it is possible to create test procedures to check the performance of a given system against a benchmark. Not only the protocol elements can be tested for correctness and compliance with the standard, but the procedures related to timing, error handling, and others can be tested. In the case of the SCAI standard, the only testing that can be performed is of the protocol elements. A protocol element, for example, representing a Call_Clear function, can be tested for correctness and compliance. The procedures associated with the handling of any functions are impossible to test because they are dependent on many factors outside of the control of the SCAI standard. Perhaps an application that uses SCAI functions can be tested, but this is considered service testing, not protocol testing, and must be performed for each application separately.

In simple terms, conformance to the SCAI standard requires the following:

1. Use of some of the functions defined in the SCAI standard
2. Use of the protocol as defined in the SCAI standard for the functions in (1)
3. Use of ROSE and ACSE as defined in the standard
4. Use of the procedures associated with the functions in (1) as defined in the standard

Chapter

6

SCAI Applications

The SCAI standard provides a mechanism for exchanging information between telecommunications and data processing environments. The basic functions available in the standard right now are coordination and third-party call control. These are high-level functions and there can be a very large set of services requiring them, but not all services would use the functions the same way. The developers of the SCAI standard recognized that a potential functional set that can support many services would be very large and, therefore, they decided to focus on the services (applications) that are most urgent. The urgency was gauged by the marketing departments of the companies involved with the standard development and by the industry groups that are charged with the development of the market-oriented applications. Two industry groups provided input into the T1S1 decision-making process regarding selection of applications requiring support by the SCAI standard:

1. North American ISDN Users' Forum (NIUF)
2. Information Industry Liaison Committee (IILC)

The NIUF develops industry/users agreements on the implementations of ISDN services. These agreements are initially based on available and developing standards, but extend beyond the standards. The agreements focus on all issues necessary for the successful introduction of services in the marketplace. One important characteristic of NIUF is the participation by users. Among the NIUF participants are banks, insurance companies, universities, and other such institutions. These users impact the agreements by participating in the deliberations. It is important because standards committees are typically not heavily attended by users, although they can attend. NIUF, while focusing on ISDN services, is a premier United States forum for discussing applications and its input to T1S1 regarding SCAI applications was considered extremely important.

The IILC is working on the issues of open network architecture (ONA) mandated by the FCC. Its participants are manufacturers, telephone companies, and enhanced service providers (ESPs). The ESPs can, to a certain degree, be considered users because they use the technology to provide services to their clients, and, at the same time, they are aware of their clients' needs.

Because of users' participation in IILC and NIUF, the input these organizations provided to T1S1 was based on the real needs of real users. T1S1 developed an initial list of SCAI applications consisting of several items. Input from IILC and NIUF clarified the needs and the decision was made to concentrate on two high-priority applications: inbound telemarketing and outbound telemarketing.

Since these applications are implemented with a degree of difference, depending on the local requirements there were variations in opinion regarding the inclusion in the standard of some functions. Nevertheless, different manufacturers support different clients with different needs and it was desirable for the standard to include at least some of the optional functions. At the same time it can be said that, although the standard is specifically defined to support two high-priority applications, the functions that support these applications could also support other applications with no changes in the protocol.

There is another point worth mentioning here, regarding application names. Sometimes the same applications have different names (depending on their creators), and sometimes it is not clear what is meant by the application name. In our discussion of SCAI applications we will use some descriptions as they were used in the SCAI standard development, and others as they can be implemented from my point of view because no sufficient discussion took place in the committee.

6.1 Inbound Telemarketing

Inbound telemarketing is an application that monitors incoming calls, receives information about the caller, possibly directs the call to a specific telemarketing agent, delivers a data screen to the agent simultaneously with the voice call, and allows manipulations of the call by the agent (transfer, conference, etc.). As its name implies, inbound telemarketing is used by telemarketing firms that sell products over the telephone after their products were advertised through media, such as television, radio, and billboards.

Figure 6.1 shows a generic arrangement for inbound telemarketing. For simplicity, only one agent and one customer are shown, although in reality there are many agents and customers involved. For SCAI pur-

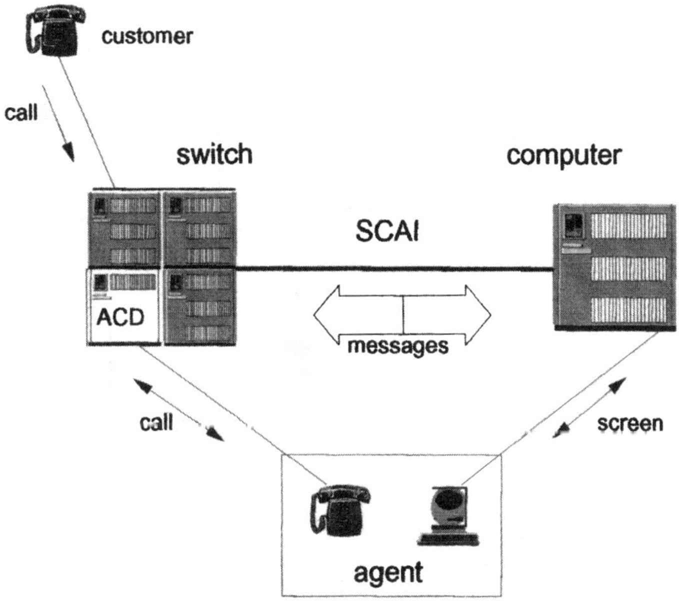

Figure 6.1 Inbound telemarketing.

poses, ACD is treated as part of a switch and its functions are not distinguished from other switch functions.

Depending on the particular requirements of the organization installing SCAI technology, the application may be focused on different functions. If telemarketing agents are equipped with headsets, they could not invoke any functions through the switch and would use computer-invoked functions such as transfer. If the setup of the particular telemarketing organization is such that no conferences are allowed, then the conference function is useless. If the setup requires transfer of calls to supervisors any time a customer asks for it, the system must be engineered to handle many transfer requests. These are examples of issues that should be considered when SCAI equipment is being designed and installed.

The variety of alternatives for an application makes it difficult to present the exact and complete flow of messages for an application because any attempt to generalize the flow would fail. However, there are some generic exchanges that take place when an application is running. Let's start developing a message flow for a generic inbound telemarketing application.

Any telemarketing application starts with agents signing in and getting ready to receive calls. When a call arrives at a switch, it transi-

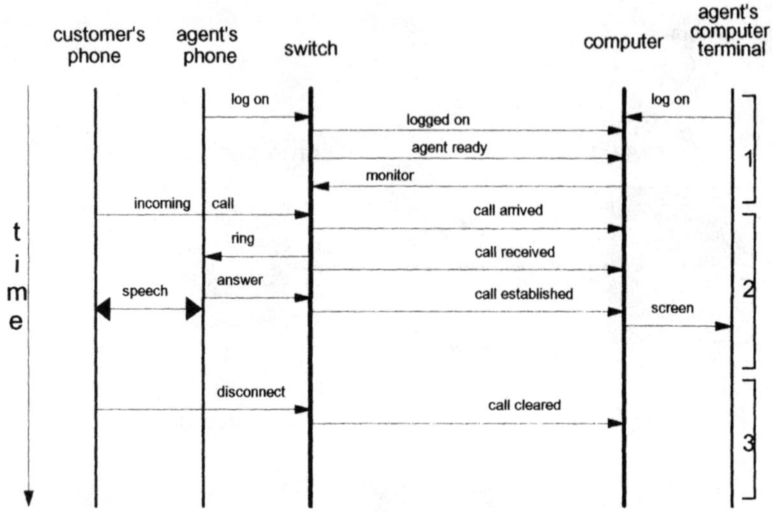

Figure 6.2 Simple inbound telemarketing.

tions into a specific call state; the state of a call changes with time. A switch sends reports on the state changes to a computer. A computer sends requests to a switch to perform specific actions, or the actions are requested by an agent via the agent's telephone connected to a switch and then the switch reports the call transitions to the computer. These are the general interactions between a switch and a computer in the case of inbound telemarketing. Figure 6.2 shows the generic part of information exchange as a function of time. The interactions in Fig. 6.2 can be divided into three classes.

1. *Initialization.* The following interactions occur:
   a. An agent logs on to a computer through the computer terminal and on to a switch* through a telephone. How the switch becomes aware of the agent's readiness is not discussed in SCAI; the important point is that the switch gets this information.
   b. While the agent progresses through the log-on procedure, the switch reports to the computer the states of an agent. The "logged on" message indicates to a computer that an agent is getting ready to receive calls.
   c. The "agent ready" message indicates that the agent is ready.

---

* Usually agents are connected to an ACD, but for the purposes of SCAI an ACD is part of a switch and we can use the terminology of associating agents with switches.

d. The messages in 2 and 3 contain the device identifier the agent is at, and this information is used by a computer to issue a "monitor request" message that sets up a monitor for the agent's device* (telephone). Now, the events that the application wants to observe at this device will be reported. The variation of the monitor setup is a presubscribed monitor: the monitoring setup is part of a switch setup. In this case, a "monitor" message is not sent.

2. *Call arrival.* When a call arrives at a switch, the following interactions take place:
   a. The call just arrived and a call ID was assigned to it; this call transitions to the ARRIVED state, and this fact is reported to a computer in a "call arrived" message.
   b. A switch is in the process of identifying the terminating device for the call. When the device is identified, it is alerted to the incoming call through the application of a ring; at the same time a "call received" message is sent to the computer to indicate the transition of the call to the RECEIVED state.
   c. When the agent at the selected device answers, several actions occur, for practical purposes, simultaneously:
      - The call is cut through to the agent and the customer and the agent and engaged in a conversation; this is represented in the figures by the arrow labeled "speech."
      - A "call established" message is sent to the computer to indicate call completion.
      - A screen with information (data) pertaining to the caller is issued to the agent's computer terminal; the caller's identification information, which permitted selection of appropriate computer files, was received in a "call arrived" message as the "calling device" parameter. The identification can be Automatic Number Identification (ANI) supplied by a modern signaling system or another form of caller's identification. If an automatic means for identification is not available, there are other methods for obtaining the ID reference. For example, a voice response unit can be attached before the call progress continues; the voice response unit collects additional information from the caller (customer ID, social security number, etc.) and passes it to a computer. Another issue with the screen informa-

* One convention used in this chapter: a "monitor" message is issued for as many devices as we discuss. In some further examples of applications, when more than one agent is involved, the "monitor request" is assumed to be issued to monitor all involved devices.

tion delivery is its timing; I show it as being performed after the call was established, but it can be initiated after a call is received or after a call has arrived, if no queuing is required. The goal of coordination is to make the customer's information delivery simultaneous with the voice call delivery; however, computers work a lot faster than people can react and small deviations in synchronization of voice calls and data delivery are easily tolerated. In any case, there is currently no provision in SCAI for a totally deterministic delivery of data simultaneously with voice calls; when such a function was proposed, the prevailing opinion was that, unless a computer takes an extraordinary amount of time for data searches (e.g., several databases have to be involved), there is no need for a specific deterministic data delivery function in SCAI. (Practical experience will tell us who is right!) In the meantime, an implementor decides when to trigger delivery of the data screen according to experience with a particular system.

3. *Call disconnect.* When the call is disconnected by the customer, as in Fig. 6.2, a switch clears the resources associated with the call and issues a "call cleared" message to the computer. There are two more alternatives for call termination. The first is when an agent terminates a call via a telephone. In this case, the same "call cleared" message is sent. The second is when an agent terminates a call via a computer terminal. In this case, a computer issues a "clear call" request to a switch; the switch disconnects the parties engaged in the call and sends a "call cleared" message.

We considered only one variation of the generic exchange of information that takes place when the inbound telemarketing application is running. Another variation includes the use of routing functionality to deliver a voice call to a particular agent selected not by the switch, but by the computer application. Let's examine Fig. 6.3.

The use of the routing function in Fig. 6.3 triggers two additional interactions between a switch and a computer: "route request" from the switch and "route selected" back to the switch. All other activities are the same as in Fig. 6.2. Note that the screen is delivered at a different place in the sequence of activities (compared with Fig. 6.2), but, as we discussed before, triggering of the screen delivery is an implementation decision and is based on the experience of an implementor.

Now I would like to present one more variation, this time related to the handling of the agents. In the interactions shown in Fig. 6.2 the agent involved with a call stays in the READY state while accepting calls in sequence. Sometimes, however, an agent, although not participating in the voice call any longer, needs to complete some work

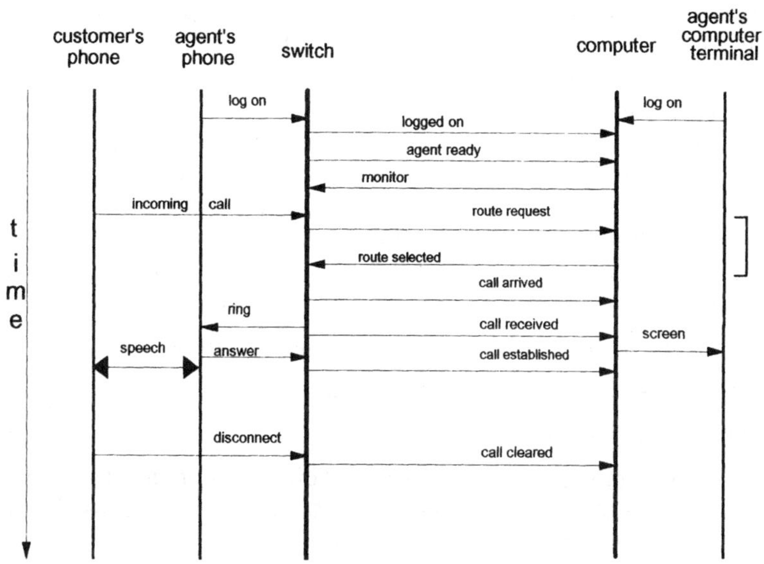

Figure 6.3 Inbound telemarketing with routing.

related to the last call. In some situations this agent must be available to receive new calls (as in Fig. 6.2), and in other situations it may be more productive to allow the agent to complete all work before receiving a new call. The latter can be accomplished with SCAI and the interactions in this case would appear as in Fig. 6.4.

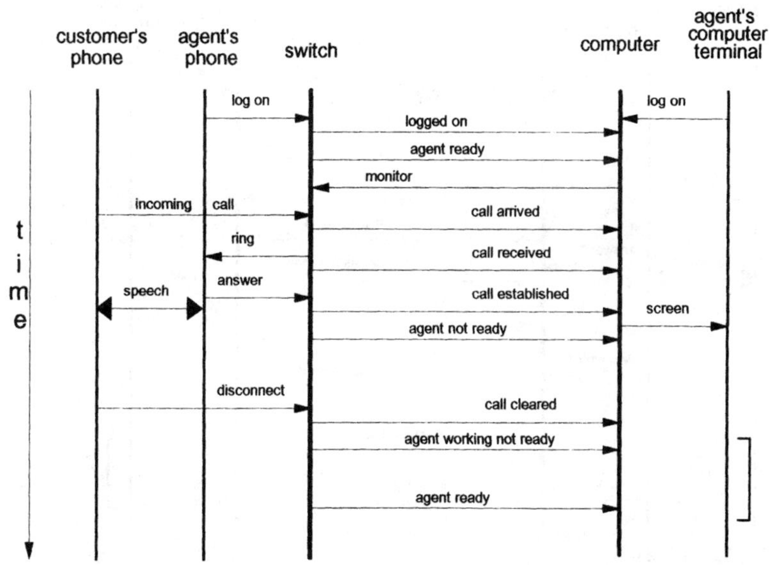

Figure 6.4 Agent-related variation.

"Agent working not ready" indicates that the agent is busy and cannot be disturbed. When the agent finishes and can again be available to receive new calls, an "agent ready" message is sent.

There can be additional functions used in the inbound telemarketing application, but it is more appropriate to discuss them as part of the customer service center application. Here, I would like to complete the introduction of interactions between a switch and a computer with the agents' signing-off procedure shown in Fig. 6.5.

When an agent signs off the system because the work day is over or it is lunch time, or whatever, the following interactions occur:

1. An "agent not ready" message is sent to the computer.
2. The agent logs off the switch and computer.
3. An "agent logged off" message is sent.

The procedures of signing on and signing off by agents are the same for all applications that require agents. They will not be discussed any more with the other applications.

The inbound telemarketing application benefits customers, users, and organizations that deploy it. Customers benefit from a better, faster service. Agents use fewer keystrokes to complete transactions and also benefit from a friendlier environment. As a result, everybody experiences less frustration. Organizations benefit from better customer relations, from a better working climate generated by more satisfied employees, and from the direct cost savings that automation brings.

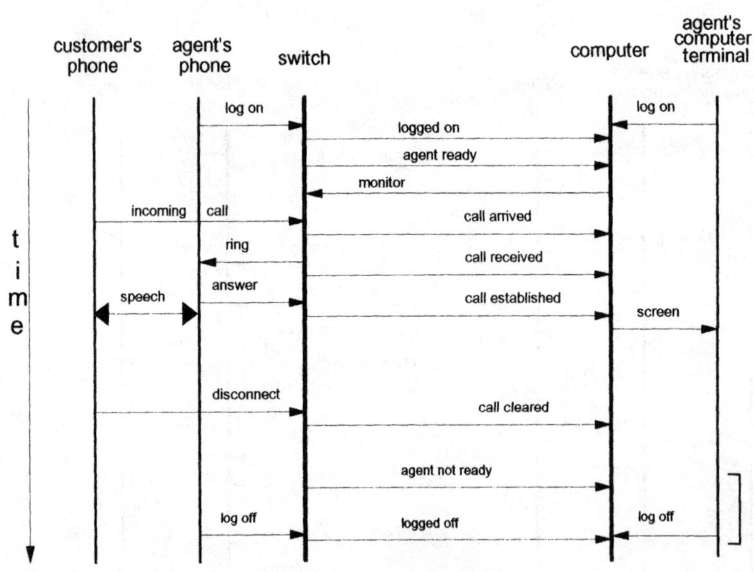

Figure 6.5 Agent log off.

6.2 Outbound Telemarketing

Outbound telemarketing is an application that makes outgoing calls to customers, connects customers with available agents, and delivers information screens to agents. The application is used by telemarketing firms that sell their products over the telephone and sales are originated by the firms themselves, not by potential customers, as in the case of inbound telemarketing. While the goal of both types of telemarketing applications is the same—to sell products—the applications are engineered differently because one deals with incoming calls (inbound), and the other with outgoing calls (outbound). In terms of the SCAI call model (remember that abstraction?!), the inbound telemarketing applications work with terminating calls (the terminating call model), and the outbound telemarketing applications work with originating calls (the originating call model). Hence, the call state transitions seen by two types of applications are completely different. Figure 6.6 shows a generic arrangement for outbound telemarketing.

Two alternatives exist for generating outgoing calls in SCAI: (1) using a straightforward make_call procedure and (2) using a predictive_make_call procedure. Consider the generic part of outbound telemarketing with make_call shown in Fig. 6.7 (the agents' log-on and log-off procedures are not shown).

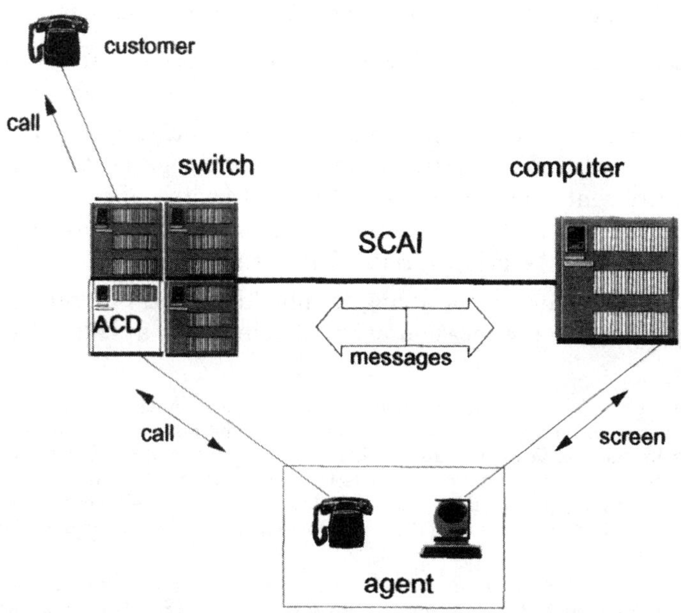

Figure 6.6 Outbound telemarketing.

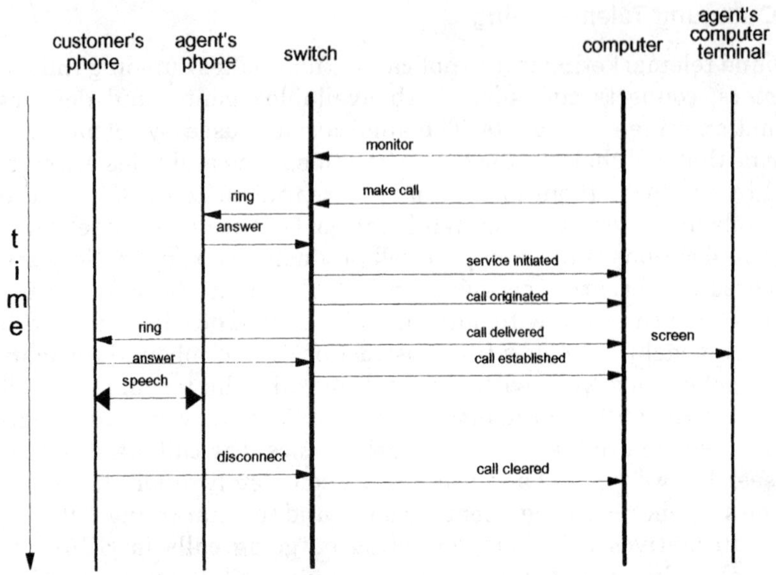

Figure 6.7 Outbound telemarketing with make_call.

After monitoring of the appropriate agents' devices is established, a computer issues a "make call" message to a switch. The message contains the intended destination identification (e.g., a directory number) and the intended agent identification. The switch does the following:

1. The intended agent is rung and the answer is expected* (and received).
2. An outgoing call is initiated on behalf of the intended agent. The destination number is "dialed" by the switch (digits representing the destination are sent over the trunk to the next switch, if the destination is remotely located). This activity is reported in two messages sent to the computer: "service initiated" and "call originated."
3. The destination is alerted via a ring (applied by a local or remote switch, depending on the location of the destination†; a "call deliv-

* Ring may not have to be applied in this situation; the application designed in close cooperation with the ACD portion of the switch may assume the responsibility for the agent's readiness to participate in the call. In this case no agent-related ring/answer actions are performed and the switch just connects the agent to the intended outgoing call. Similar consideration applies to the agent-related ring/answer in the inbound telemarketing application. Showing this activity in the message flow diagram makes the diagram more general.

† Remember, the SCAI switch represents the entire telecommunications network "hidden" behind it, and the actions of the network are represented to the SCAI computer as the actions of the switch.

ered" message informs the computer about the destination alerting, and the computer delivers the customer's information screen to the agent's computer terminal (again, the screen delivery may happen in various places. This action after "call delivered" seems more natural to me, but it completely depends on the application's requirements and performance of the particular equipment).

4. When the destination answers, the following activity takes place simultaneously:
   - The call is completed and the agent and the customer are engaged in the conversation.
   - A "call established" message is sent to the computer.
5. When the destination disconnects the call, "call cleared" is sent to the computer.

As we discussed before, disconnect can also be initiated by the agent, and the state of the agent after the disconnect can be either READY, NOT READY, WORKING READY, or WORKING NOT READY.

Now, let's consider what the difference is in the message flow if predictive dialing is used. Figure 6.8 shows the interactions.

The difference between regular and predictive dialing is the sequence in which a switch connects one party with another. In the case of regular dialing using a "make call," the agent is connected first

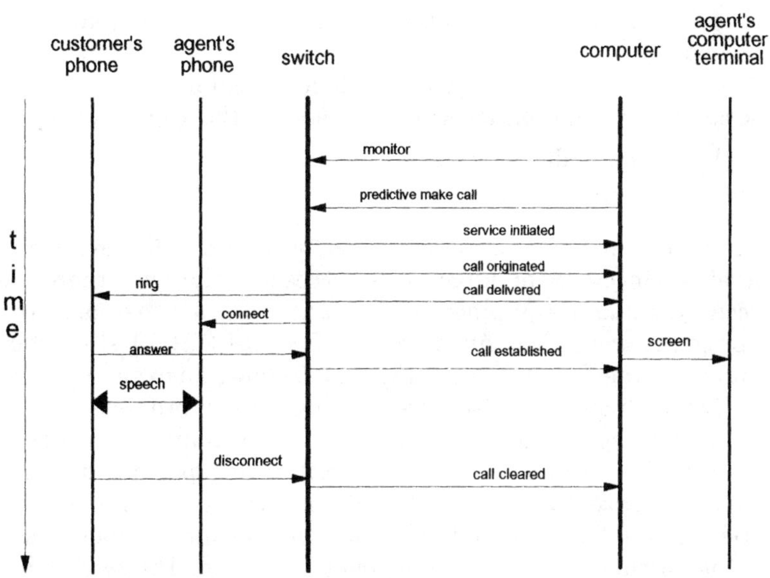

Figure 6.8 Outbound telemarketing with predictive dialing.

and then the customer, if answered. In the case of predictive dialing, the agent is connected only when the customer is either found (as in the example in Fig. 6.8) or when the customer actually answers. Predictive dialing saves resources because there are no agents that are idle while waiting for a customer to answer. It is achieved by simultaneously placing many calls to potential customers and connecting agents after customers' answers or when calls are almost completed.

Two types of predictive dialing are available in SCAI: (1) an agent is connected in the DELIVERED state (destination is alerting) and (2) an agent is connected in the ESTABLISHED state (destination has answered). Figure 6.8 presents option (1); in the case of option (2), the action "connect" would follow the "call established" message. Here is the sequence of events when option (1) predictive dialing is used:

1. After monitoring of appropriate devices is established, a computer issues a "predictive make call" message which carries the destination number (customer) and the origination number (agent); a switch originates a call on behalf of the agent.
2. The progress of the call is indicated to the computer via "service initiated," "call originated," and "call delivered" messages, the last message being sent after the destination has been alerted (ring).
3. A switch connects an agent.
4. When the destination answers, the following actions are performed simultaneously:
   - The voice call is complete and the agent and the customer are engaged in the conversation.
   - A "call established" message is sent to a computer.
   - The customer's information screen is issued to the agent's terminal.
5. When the customer disconnects, a "call cleared" message is sent to a computer.

So far, in both applications' descriptions, we assume that every call attempt, either incoming or outgoing, is successful. However, there are faults occurring due to equipment or human error. In either case, the call in progress would not finish in the ESTABLISHED state, but would end up in the FAILED state. Figure 6.9 shows an example of the message flow with an error discovered during the call progress.

After a switch receives a "make call" message, it connects the agent (ring/answer) and initiates an outbound call to the specified destination. Two reports are issued while the call progress is continuing: "service initiated" and "call originated." When the network cannot find a route to the destination, the SCAI switch is informed. The switch discontinues call progress for this call and issues a "call failed" message to

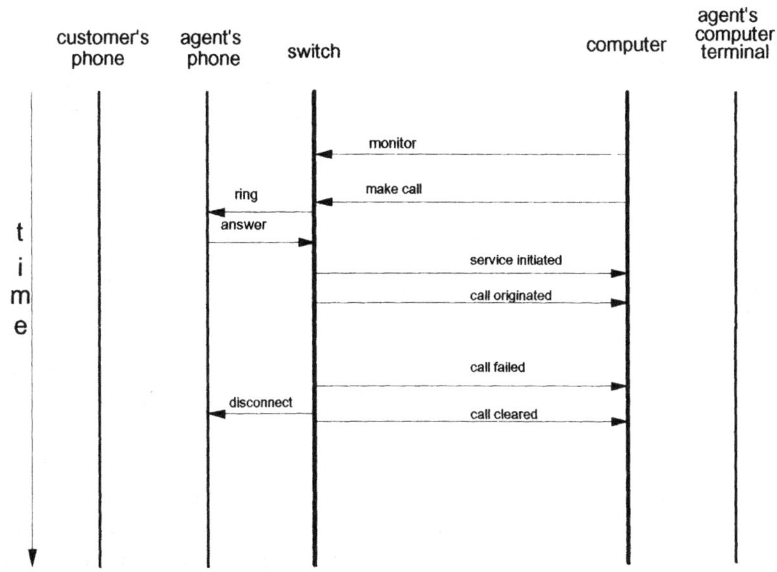

Figure 6.9 Failed outbound call.

the computer indicating that the call is now in the FAILED state. The cause for the transition included in the "call failed" message is "no route to destination." Then the switch frees the agent from the attempted call (it is shown as the arrow labeled "disconnect") and terminates the call. The "call cleared" event informs the computer about the call termination.

As in the case of inbound telemarketing, the outbound telemarketing application benefits customers, agents, and organizations due to faster service, a better working environment, and savings. Of course, a potential customer may not see it that way if disturbed at dinner time.

6.3 Customer Service Center

The Customer Service Center is the application that supports the activity of a customer service department. Examples of these applications are many and easy to explain because they are so familiar to us. Just think of times when you call a bank, an insurance company, or a telephone company. We know the type of conversation that usually takes place. What we do not know is what kind of constraints are imposed on a particular service department by its management. The same constraints are important to recognize when a SCAI customer service application is being designed.

The customer service center application is similar to inbound telemarketing (see Fig. 6.1). The difference is not in the technique, but in the emphasis. While inbound telemarketing applications deal with the situations when a customer is willingly calling to purchase a product, in the case of a customer service center, customers call either with complaints or questions. Because of the more intensive environment of the customer service center, the use of supplementary services like *hold/retrieve* and *transfer* is greater than with telemarketing. When a customer service center call does not require any supplementary services, the message flow is exactly as in the case of inbound telemarketing. Hence, instead of discussing the issues already covered, I would like to consider what happens when supplementary services are used. Figure 6.10 shows an example of an incoming call with a hold/retrieve function applied.

Portions 1 and 2 of Fig. 6.10 describe an agent log-on procedure and the establishment of a call between an agent and a customer. "Speech 1" signifies a conversation between the agent and the customer. Portion 3 shows what happens when the agent decides that a consultation with a supervisor is necessary and wants to place the customer on hold. Here are the interactions:

1. The agent requests a hold via the computer terminal.
2. A computer issues a "hold call" request to a switch.

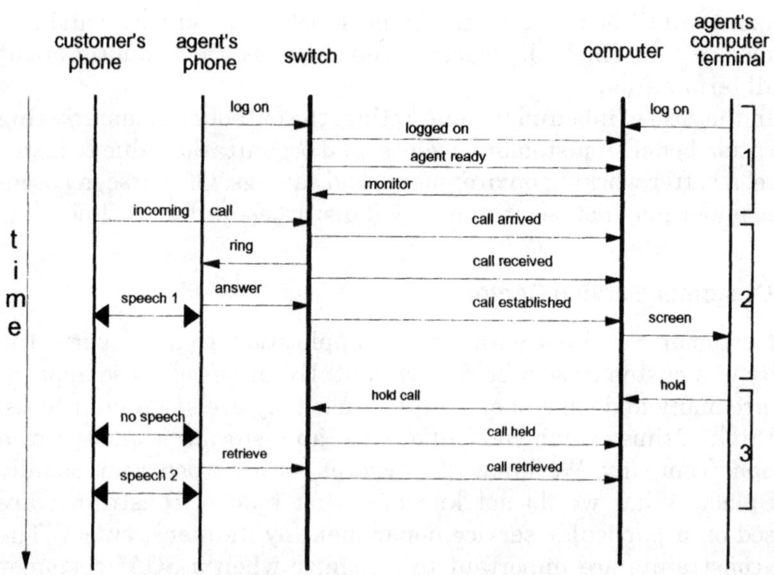

Figure 6.10 Hold/Retrieve flow.

3. A switch places the customer on hold and no conversation can take place now (no speech); then the switch reports the actions to the computer via a "call held" message.

When the agent is finished talking with a supervisor, the following interactions take place:

1. The retrieve request is generated via the telephone (not via computer this time, just to present both alternatives).
2. The switch retrieves "the customer" and the conversation continues (speech 2).
3. The switch informs the computer via a "call retrieved" message.

Another widely used feature in the customer service environment is *transfer*. Transferring of customers to other agents or to supervisors is very common for many reasons. A customer may insist on speaking with a supervisor because he or she is unhappy with the results achieved so far; an agent may decide to transfer a customer to a supervisor because only a supervisor can resolve a particular problem; an agent may have to transfer a customer to another agent who is more suited to answer a question. In all these cases, two transfers have to take place: transfer of a voice call and transfer of associated customer information (data), so the dialog can continue from where it left off with a previous agent. SCAI allows this to be accomplished. (See Fig. 6.11.)

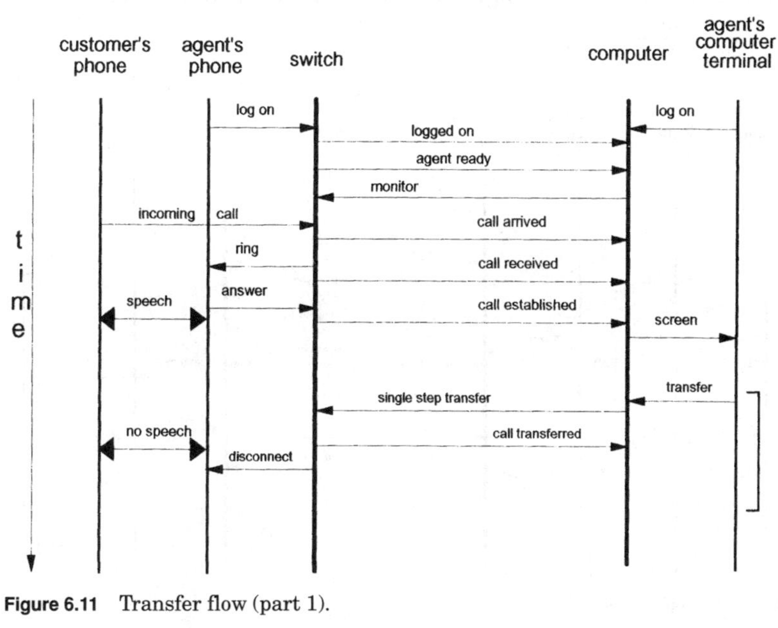

Figure 6.11 Transfer flow (part 1).

Figure 6.11 shows an example of a single-step transfer. With this type of transfer, a call is being transferred from one agent to another, and the transferring agent would be free to receive other calls. The type of diagram that is used here to show the flow of messages and actions is not quite appropriate for functions that involve more than two participants in a call. Another diagram can be used for this purpose, but first we should examine Fig. 6.11 because it does show the interactions from the point of view of a transferring agent.

When an agent decides to transfer a call to another agent, he or she generates a request for transfer at the computer terminal and a computer sends the request in a "single-step transfer" message. In addition, a computer transfers the customer's information screen to the same destination as that of the voice call. A switch transfers the call to the destination indicated in the request message, sends a report on this action to a computer in a "call transferred" message, and disconnects the transferring agent from the call, allowing the agent to be available for other calls. From the point of view of the transferring agent, the call does not exist any more. For the agent who deals with this transferred call, it is a new call.

Figure 6.12 shows the same single-step transfer using a different type of diagram. In this diagram, two agents are represented by their computer terminals and telephones: agent 1 and agent 2. Agent 1 was handling the call before the transfer (portion 1) as an incoming call. What occurs when agent 1 decides to transfer the call to agent 2 is shown in portion 2 of the diagram. The interactions are as follows:

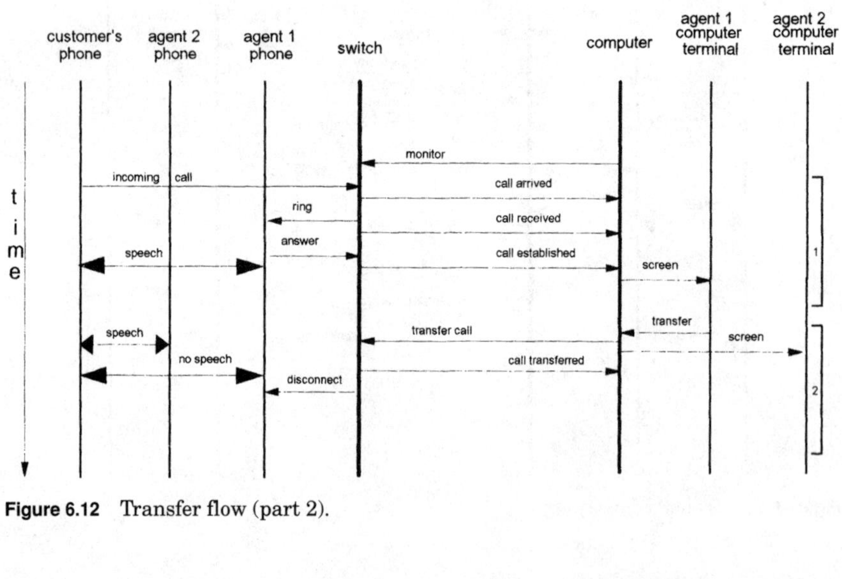

Figure 6.12 Transfer flow (part 2).

1. Agent 1 generates a request for transfer at the computer terminal.
2. A computer formulates the transfer request in a "single-step transfer" message to a switch; the intended destination is included. A computer also delivers the same customer's information screen to agent 2.
3. The switch performs several actions:
   - The call is transferred to agent 2 (it is indicated by an arrow between the customer and agent 2 labeled "speech").
   - Agent 1 is disconnected from the call (notice that there is no speech path between the customer and agent 1).
   - A "call transferred" message is sent to the computer.

Another function is very useful for all applications discussed so far. This function is *filter,* and it changes the types of reports that are being sent to a computer for a monitored device. Figure 6.13 shows the interactions when filter is used.

The example in Fig. 6.13 shows that initially "monitor (all)" was sent to a switch to indicate that all events at the specified device are to be reported. When a call arrives, all call state transitions are reported to a computer (call arrived, call received, call established). The dashed line indicates that a number of calls were handled by our agent and that the agent is again available to receive new calls. However, before the next call is assigned to the agent, a "change filter" message is sent

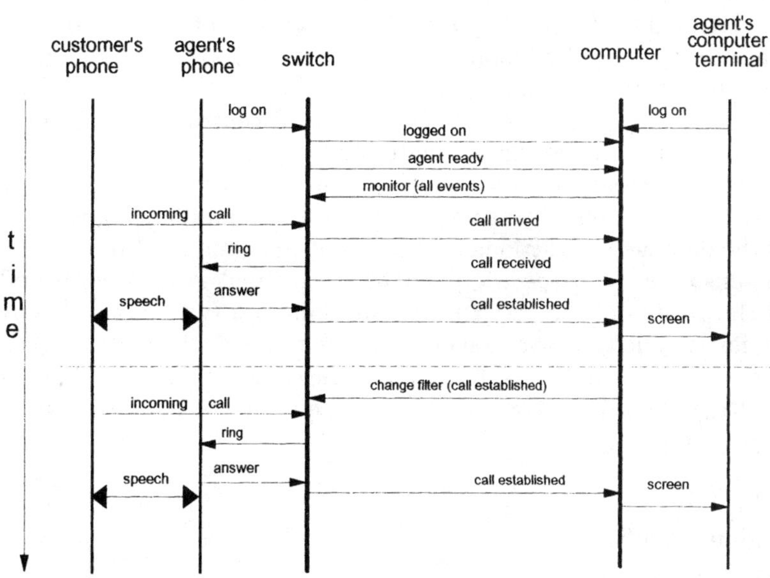

Figure 6.13 Filtered flow.

to a switch which indicates that only the "call established" event is to be reported. The application may be so busy with other actions, that it does not have the time to analyze every detail of call processing information and can operate only on the knowledge of calls being completed (established). When a new call arrives, the switch processes it as before, but the computer receives only a "call established" event report. At this time, the customer's information screen can be issued.

While the customer service center application is very similar to the inbound telemarketing application, it is usually more intensive. Often, customers call with complaints and problems. The friendly environment that SCAI helps to maintain assists the agents in providing a faster and more personalized service to their customers. In the cases of transfers, the customers do not have to repeat the same information. The result is a better working environment for everybody.

6.4 Emergency

Emergency is a generic name for the types of applications that report critical information and can help alert appropriate entities (live or otherwise) of the occurring problem. The simplest example is the well-known *911 service.* With this service, a caller to a police station is immediately identified by the calling number, which is translated into the caller's location. More elaborate examples of emergency applications are an unmanned fire alarm system, industrial control system, or an observation system. There can be a vast number of different applications fitting into these categories. As an example of the actions taking place in these applications, we can examine the industrial control system shown in Fig. 6.14.

The emergency notification for an industrial control system uses a SCAI link to inform the plant's manager, who is at home after work hours, of the problem. The problem was discovered by an unmanned control system consisting of sensors connected to a computer. Figure 6.15 shows the detailed interactions in the case of emergency notification.

The sensors control specific parts of the technological process and report the results of the activities to the computer (we show only one sensor, for simplicity). When no faults are found with the technological process, an indication of "good" is sent to the computer. When a fault is discovered, the failure report is sent to the computer with an indication of the following:

- Type of fault
- Location of fault
- Time of occurrence

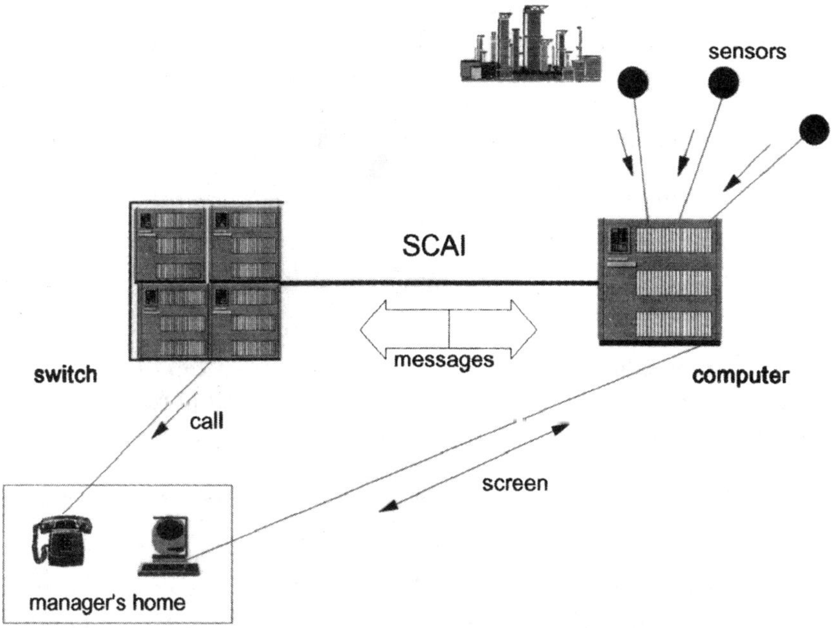

Figure 6.14 Industrial control system.

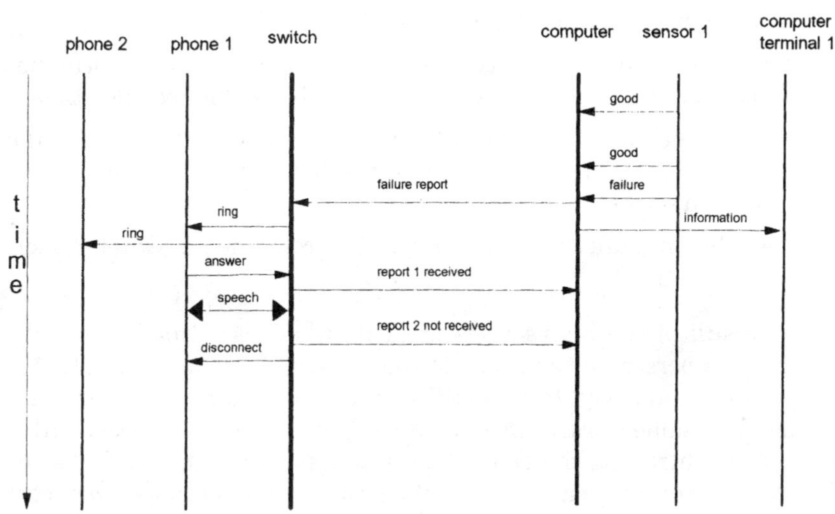

Figure 6.15 Emergency notification.

The computer analyzes the information and identifies the procedure to be followed in the case of this particular fault occurrence. According to the procedure, two people have to be notified. The computer formulates a message to the switch (which is connected to the computer over a SCAI link) containing the following information:

- Telephone number of the first person to call
- Telephone number of the second person to call
- Address of the announcement system to be attached to the call when a called person answers

At the same time, the computer issues a detailed report on the failure to the computer terminal that belongs to the first person on the list (this terminal is collocated with the telephone of the first person, e.g., at home).

A switch, after the receipt of a "failure report" message, performs the following:

- An outgoing call to person #1 is generated (to phone 1).
- An outgoing call to person #2 is generated (to phone 2).
- When person #1 answers, an announcement system is attached (it is represented in Fig. 6.15 as "speech" between phone 1 and a switch, for simplicity); the announcement system plays a message that informs person #1 of the type and timing of the problem (e.g., "problem #4 occurred at 10 PM at location ZX, severity 1").
- A "report 1 received" message is sent to the computer with information that person #1 has been alerted to the problem at the recorded time.
- After a time-out occurs for person #2's response, a switch sends "report 2 not received" to a computer indicating that person #2 was not alerted to the problem.
- When the announcement to person #1 is complete, a switch disconnects the call.

The result of the interactions presented here is a timely alert of a responsible person to the problem that occurred. In our example, two people were supposed to be notified, but only one has answered the phone. The same person also has a computer terminal linked with a computer controlling the technological process and can use the terminal to get more information about the problem and, perhaps, correct it. Using a SCAI link in this type of an environment allows any number of people to be called in any sequence and provides for contingencies in cases when someone is not available to respond to the problem.

The preceding example is based, not on the existing SCAI messages, but on the potential messages that can be easily added to the SCAI standard. Our example requires three messages: *failure report, report received,* and *report not received.* The "failure report" message could be defined as follows (in a simplified form):

```
Failure_Report
Direction        Computer---->Switch
Parameters:
  Request:
       resp_People (list)          M
       announcement                M
  Success:   NA
  Error:     no Link
```

The message informs a switch of the telephone numbers of people to call and of the announcement system that will play the recorded message with all failure-related information (time, place, severity, etc.). If the message cannot be delivered to a switch (the link is out of service), an error is returned. The computer inputs all necessary information into the announcement system to be used, or, to make it simpler, an announcement system can contain only a generic message, e.g., "there is a problem!", and to find out more about the problem, access to a computer via a terminal is required.

The other two messages could be defined just as simply. The message "report received" should only contain two parameters: "telephone number reached" and "time." The message "report not received" could be exactly the same. Of course, this is just an exercise to show that the SCAI standard permits additions of messages that are required for specific applications, and there is already an agreed methodology for defining these messages.

We considered only industrial control application in our example. Other emergency applications are either similar to it or of the 911 type. Another way of looking at this issue is to classify 911-type applications as incoming and the other type as outgoing. The incoming applications are receiving calls, and additional information is based on these calls. The outgoing emergency applications generate calls based on some criteria. The criteria can be faults in an industrial operation, fire, temperature crisis, or even inventory dropping below a certain level. When specific criteria are met, voice calls are initiated based on a predetermined algorithm. In all cases, SCAI technology speeds up the alerting of appropriate personnel who can take actions, or delivers critical information to emergency personnel a lot quicker. The result is a much greater level of emergency service for intended customers.

6.5 Data Collection

Data collection applications are the applications that use the SCAI link to collect information in a computer after the information was entered via an accessible voice device (telephone). This kind of application would be used in situations when telephones are available and easily accessible, while computer terminals are too costly or inappropriate or otherwise not suitable. A good example is a hotel/motel application.

There can be many variations of the hotel/motel application, but the one we will discuss should present the potentials. We know that in every hotel and motel, cleaning staff clean up rooms every day and report to their supervisors the state of each room they serviced. Hotel/motel management needs to know when a room is available for new guests and whether there are any problems with a room. Doing all this without SCAI requires several different types of reports and the information is not always available in a timely manner. With SCAI the following could take place (see Fig. 6.16):

1. A cleaning person cleans up a room (#134) and checks its condition for broken or missing equipment.
2. When finished, the cleaning person uses a telephone and dials ***7 1 5.**
3. A switch controlling the telephone interprets ***7** as an invocation of the SCAI hotel application and begins to construct a message to the computer.

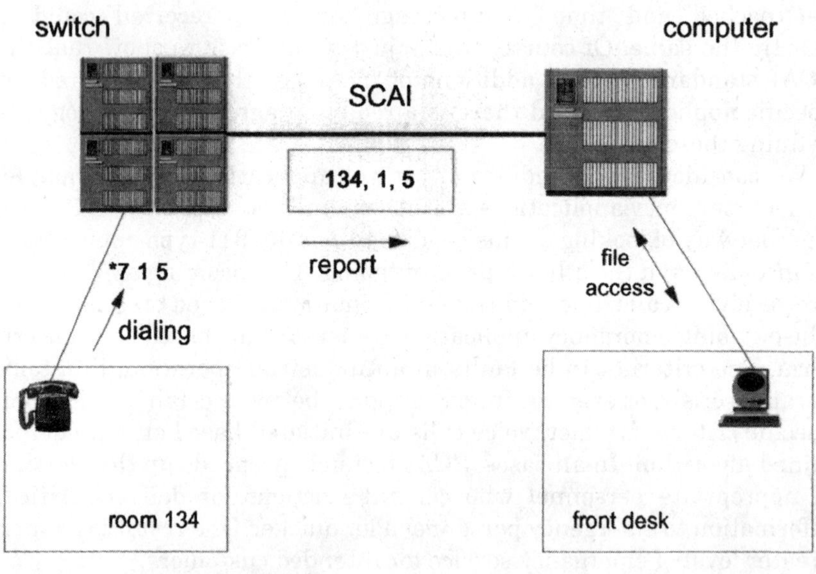

Figure 6.16 Hotel/motel application.

4. The message to the computer contains the room number (identified by the switch based on the knowledge of a line connecting the telephone to the switch), the fact that the room has been cleaned (indicated by dialed digit **1**), and the fact that the TV set in the room is broken (digit **5** dialed by the cleaning person indicates this).
5. The computer receives the message and updates the file for room #134.
6. The file can be accessed by the front desk personnel any time to find out if the room is available or not.
7. The computer issues a message to the computer terminal belonging to the maintenance department indicating that a TV set in room #134 needs repair.*

Messages required for this application are not currently part of the SCAI standard. Actually, only one message—report—is needed for the simple case we just discussed. And, again, the message is simple: it should contain several parameters representing room number, room status, and indicators for specific items, such as a broken door. More sophisticated service may require additional messages or parameters in the report message we just identified.

6.6 Message Desk

The Message Desk application is associated with attendant services. An attendant can be a secretary or a specially trained person who handles incoming calls for a designated group of people. An attendant answers calls, takes messages, distributes messages to appropriate destinations, and so on. An attendant is equipped with a device similar to a key system. Depending on the specific device used, an attendant may just record messages in a computer or invoke a multimedia capability and store a message in different forms, such as voice or e-mail. Figure 6.17 shows an example of an attendant-based message desk application.

In the example, a *caller* is trying to reach an *intended receiver,* but the intended receiver does not answer. Instead, the call is forwarded to an *attendant.* Forwarding can be set in many different ways, for example, forward on busy, always forward, and so on. An attendant receives a call simultaneously with the screen of information on the intended receiver. It occurred because when the call has been forwarded to the attendant,

* Alternatively, a message to the switch with the telephone number of a repairperson and the address of an announcement system can be sent; the switch then would call the number and, after the answer, connect the announcement system that would inform the repairperson of the problem.

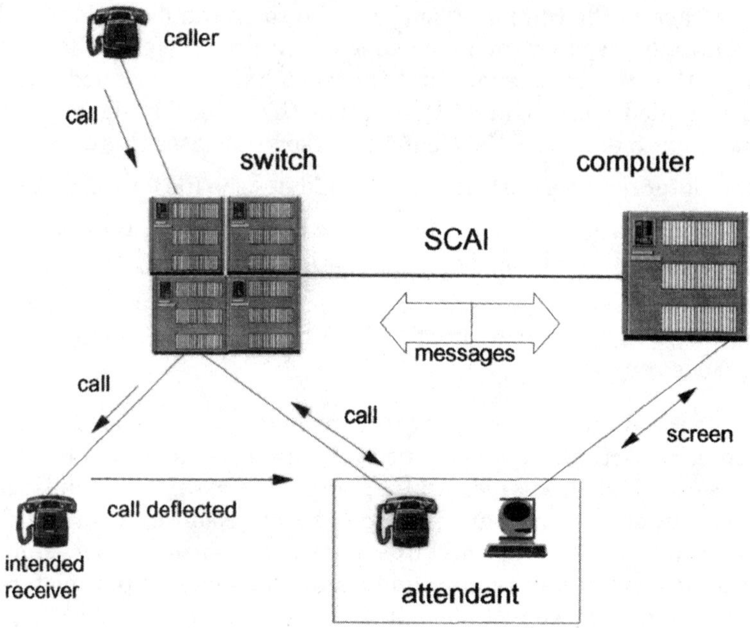

Figure 6.17 Message Desk application.

the switch "knew" that it was supposed to inform a computer regarding the initiator of this call forwarding. A message was sent to a computer, based on which the information screen was issued to the attendant. The attendant answers the call and speaks with the caller armed with the information about the party that the caller was trying to reach. This information may help the attendant to set an appointment, to find the intended party, or record a message from the caller in the format suitable for this particular intended receiver. The variations are many and totally depend on the particular rules in the organization, a particular call, and the technology used by an attendant. Let's now examine a generic message flow more closely in Fig. 6.18.

An outside caller makes a call to someone within the organization, who is an intended caller. When the switch attempts to terminate the call to the intended receiver, it recognizes that this phone is set for call forwarding (this is shown in the flow diagram as a pair "termination attempt/deflection"). The switch "knows" that the calls are to be forwarded to a number specified when forwarding was set, which, in this case, is the phone number of the attendant. The switch rings the attendant and, because it is the call destined to the attendant, sends the computer a message indicating that a call was deflected from a specified phone. This message triggers a screen of information regarding

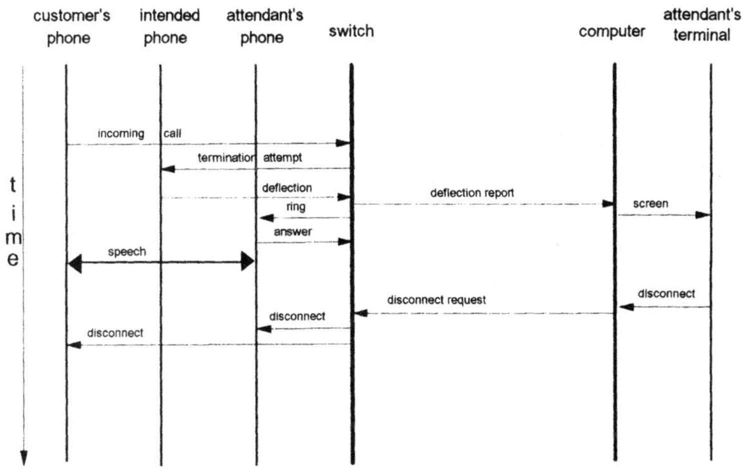

Figure 6.18 Message desk flow.

the owner of the phone to be issued to the attendant's computer terminal. When an attendant answers the deflected call, a conversation is established between a caller and an attendant (speech). During the conversation, the attendant can use information about the intended recipient of the call as appropriate. For example, the attendant can set up an appointment with the intended person based on the person's available schedule. When the conversation is over, the agent generates a disconnect request from its computer terminal,* which is sent to the switch in a "disconnect request" message. The switch disconnects the parties involved in the call.

This example of the message desk application resembles other applications that are built around incoming calls (e.g., customer service center). There are differences, though, between them. First, the message desk application, at least the variant of it discussed here as an example, does not track an incoming call as closely as the customer service center application does. The call is presented to a computer when it is established. Second, the message desk application requires the retrieval of information about the called party, that is, the destination of the call. In other incoming call management applications, the information retrieved was always about the caller (originator of the call). Now, the retrieval of information about the members of the organization does not require any special technique to be created because the

* In most cases, the disconnect would probably be originated by an outside caller; I just want to show that it would be possible to use SCAI for call control functions with the message desk application.

incoming telephone number is used as the key to record retrieval, and the incoming telephone number is always known.

Can the message desk application be supported by existing SCAI messages? Absolutely, if the version of the message desk application presented here is acceptable. Two SCAI messages could do the job: Call_Established and Clear_Call.

Call_Established can be used as a "deflection report." It contains all necessary parameters: call ID, calling device (the caller), called device (the intended receiver), answering device (the attendant). It also contains a private data parameter that can be used for specific organizational purposes, and a cause parameter, which is optional and is probably not needed in this case. The monitor ID parameter can be set for an intended receiver.

The Clear_Call message can be used as a "disconnect" message, exactly as it was designed. It contains the party ID parameter that uniquely identifies the call to be cleared. The other parameter in this message is private data.

Message desk applications benefit the organizations because they provide a better customer interface for those in the organization whose work is customer-oriented, but who cannot always be at their desks to answer calls. The applications facilitate relations with customers and, at the same time, allow an efficient and effective distribution of information regarding received calls to the appropriate members of the organization.

6.7 Other Applications

If one reads articles and listens to presentations on CTI, it is easy to see and hear a great variety of applications that CTI technology can support. It is impossible to judge every application mentioned, but in many instances the difference between application X and application Y is not technical. Many times the difference is in the specific market segment that uses an application under a specific name. There are, of course, differences in the way the same application is provisioned for different markets. There are even differences between applications engineered for different companies in the same market segment. Nevertheless, an application like customer service center is really the same application regardless of some differences in the message set used by various companies.

The SCAI standard supporting CTI technology can be used in a variety of markets. These markets can be identified as follows:

- Finance and Banking
- Health

- Transportation
- Manufacturing
- Political organizations
- Education
- Government

Sometimes it is even difficult to identify the market segment for a particular application because the application crosses these segments. For example, an application *Credit Collection* is applicable in many market segments: medical, education, retail, government, and others.

In many cases, applications that have different names are not really different and have the same basic requirements as the ones that we have already discussed. *Credit Collection, Fund Raising, Market Research,* and the like are applications designed respectively for all markets, political organizations, and market research companies. However, they all are basically the same application—outbound telemarketing, or to use a better term, outgoing call management. In all three cases the application generates calls to specific people (telephone numbers) and delivers screen information to the operators (agents) who are canvassing people for unpaid bills, political contributions, or just their opinion needed for market research.

Applications like *Specialist Referral* and *Reservations* are applicable to the medical and transportation industries, respectively, but both are really variations of the customer service center. Both applications support customers calling in with requests and questions. There are probably differences in the details, but in general the SCAI customer service center application can satisfy the needs.

The applications that the SCAI standard, and CTI technology in general, support have a very beneficial effect on the companies using them. They are versatile; that is, they can be adapted to a variety of situations. The benefits that they offer to all participants are not only financial, but also intangible, like building customers' good will and employees' satisfaction.

Chapter

7

SCAI Transport Mechanisms

The SCAI standard defines the application layer functionality and protocol and is not concerned with the layers below. However, the deployment of SCAI requires a selection of the transport mechanism suitable for a specific SCAI implementation. We define SCAI transport mechanisms as protocol stacks of layers below layer 7 that are capable of transporting SCAI application layer messages between peer systems. Figure 7.1 shows a model of transport independence, which was one of the defining principles for SCAI. According to this model, SCAI applications can use any transport mechanisms satisfying communication requirements. Potential transport mechanisms in the model are identified by protocol stacks *X, Y,* and, *Z.*

Moving away from models and closer to the reality of the transport protocols that are actually used today, we can identify a relatively small set of practical protocols. This set would consist of DSS1 (the ISDN access signaling protocol), X.25 and its sibling—Frame Relay protocol, and SS7. All protocols in this set are standards and intentionally so. Although, from the SCAI point of view, any protocol can be used that supports necessary functions, we are creating a standard way of using SCAI. Hence, the set of protocols that we identified does not include nonstandard protocols developed by various manufacturers to support their specific equipment. Let's consider each potential transport mechanism individually.

7.1 DSS1

The signaling access to ISDN is accomplished with the Digital Subscriber Signaling System Number 1 (DSS1). ISDN is a new standard way of providing telecommunications network services to users. The goal of ISDN is to integrate voice, data, and other forms of communi-

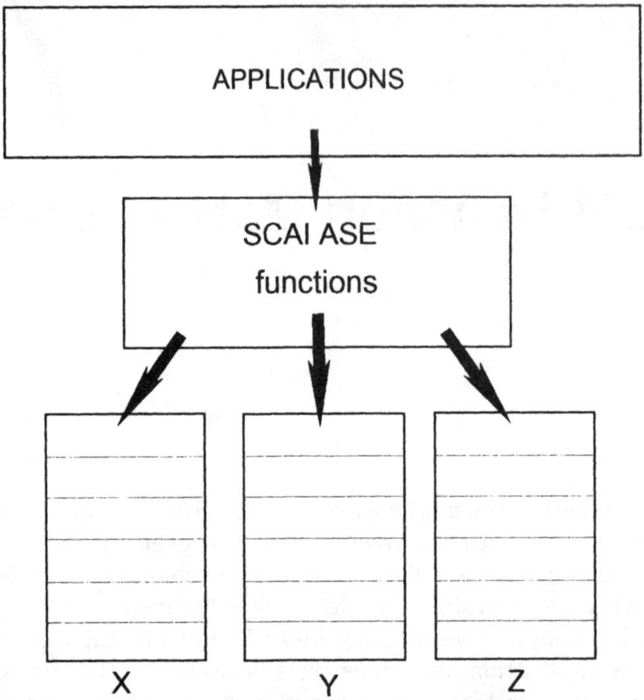

Figure 7.1 Transport independence.

cations over the same network access lines. Its promise is better, faster, more economical services that are supported by a single network. The key component of ISDN is out-of-band signaling. In analog networks, signaling is accomplished in-band; that is, information exchanged for call control is sent over the same channel as is speech (or data). Out-of-band signaling is a technique of exchanging call control information over a specific channel—the signaling channel. This type of signaling is supported by specifically designed data communication protocols. In ISDN there are two signaling protocols designed for two different elements of network communications. Figure 7.2 identifies two signaling protocols used in ISDN.

Signaling protocols on two sides of an ISDN switch are different: on the right is a D-channel signaling protocol, which is DSS1; on the left is SS7, the network (or interswitch) signaling protocol. The D-channel protocol handles user-to-network signaling; SS7 handles signaling within the network.*

* When ISDN and its signaling systems were conceived, the distinction between an access and a network seemed logical. The argument was that there are differences in requirements for those two types of interfaces. Now, with a different technological environment, the argument is not so strong, but the difference for ISDN remains and is affecting ISDN services.

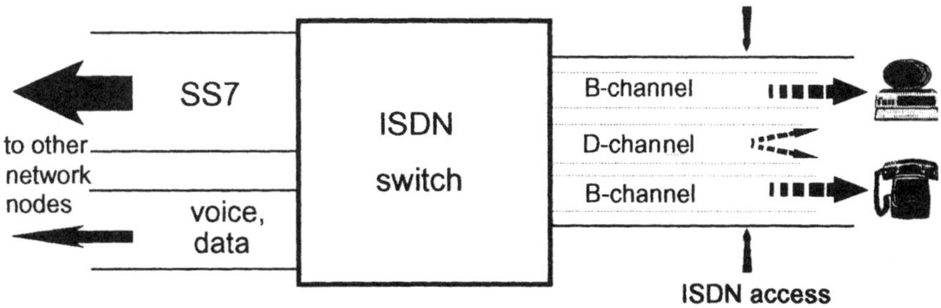

Figure 7.2 ISDN signaling.

While both signaling protocols are essentially data communication protocols that carry messages with call control (and non-call control) information between appropriate network entities, their design is not the same, except for the fact that both protocols are layered. We are not going to discuss all the details of these protocols, but more clarification of Fig. 7.2 is in order.

There are two types of ISDN access interfaces: primary rate (PRI) and basic rate (BRI). PRI consists of 24 channels, each one 64 kbps. Twenty-three channels are used for actual information exchange (voice, data, etc.) and are called B-channels. The twenty-fourth channel is used for signaling and is called a D-channel. The usual way of referring to PRI's structure is 23B + D* BRI consists of three channels: two B-channels (64 kbps each) and a D-channel (16 kbps). BRI's structure is referred to as 2B + D. In principle, there is no difference between D-channel protocols in PRI and BRI; that is, the same protocol elements exist in both. BRI is thought of as a connection to a desktop, with one B-channel "connected" to a voice device and another B-channel "connected" to a data communication device. DSS1, running in the D-channel, supports both B-channels.† PRI is usually used with T1 interfaces and is thought of as a connection to PBXs, although it can be used in any scenario requiring the bandwidth of PRI. Figure 7.2 shows BRI as an example of the ISDN access mechanism.‡

When calls are originated in ISDN, the D-channel carries call request information to the local switch, which, in most cases, proceeds with sending call request information to another switch in the network, and so on until the terminating switch is encountered. SS7 carries signaling information between various switches and other network entities involved with signaling. D-channel signaling and SS7

* The number of channels in the American ISDN standard is 24. The European ISDN standard contains 32 channels (30B + 2D).

† The B-channels of BRI can be used for any purposes. I just outlined a more traditional alternative.

‡ We will discuss more of the SS7 aspects of Fig. 7.2 in Sec. 7.2.

interwork in local switches and together provide a mechanism for ISDN call control. A general network picture with both signaling systems would be as in Fig. 7.3.

The D-channel signaling protocol, DSS1, is defined in three layers (see Fig. 7.4).

Layers 1 and 2 correspond to the OSI layers 1 and 2; they perform physical and data link control functions, respectively.

Layer 3 is different. The OSI network layer performs addressing and routing functions, and DSS1 layer 3, although it performs traditional layer 3 functions, is responsible for call control. Call control functionality includes exchange of information required to set up, progress, and

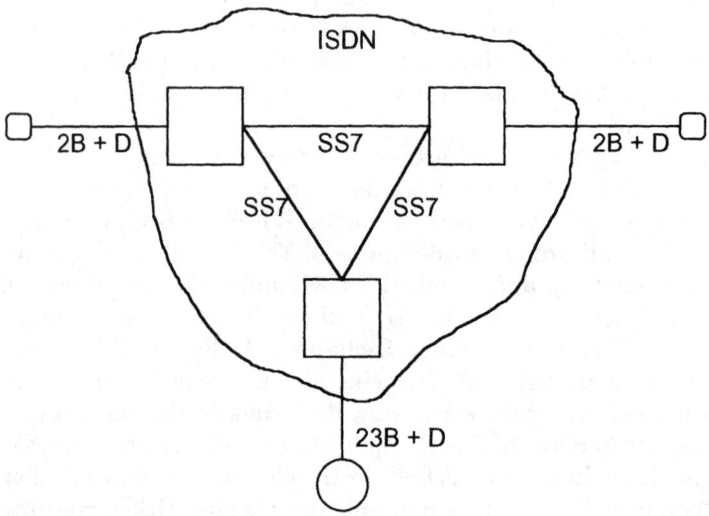

Figure 7.3 ISDN.

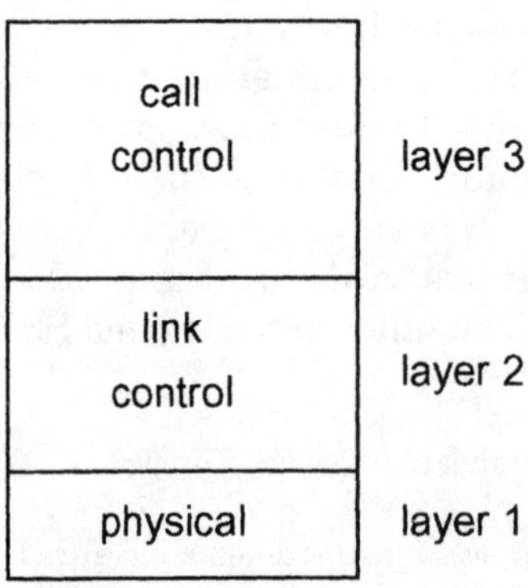

Figure 7.4 D-channel protocol architecture.

Information element	Direction	Type	Length*
Protocol discriminator	both	M	1
Call reference	both	M	2
Message type	both	M	1
Bearer capability	both	O	4–12
Channel identification	both	O	2
Progress indicator	both	O	2–4
High layer compatibility	both	O	2–5
Display	n--->u	O	special note†

M = mandatory; O = optional; n--->u = network to user.
* The length is indicated in octets.
† According to ITU-T recommendation Q.931, this length can be minimum 2 octets and maximum 34 or 82 octets, depending on the network.

Figure 7.5 CALL PROCEEDING message.

terminate a call. Layer 3 contains a set of messages used for call control, each message conveying specific information used by switches and terminal equipment to perform specific functions.* When we consider DSS1 protocol as a transport mechanism for SCAI, it is layer 3 of the protocol that we need to investigate; layers 1 and 2 would be used by layer 3 in the standard way.

Layer 3 contains two categories of messages: (1) messages for basic call control and (2) messages for supplementary services. The goal is to find a way to use DSS1 which would give the best results and would not require big changes.

The messages in category (1) are used for establishing and releasing calls, and to carry additional information pertaining to basic calls. The messages in this class are *ALERTING, SETUP, PROGRESS, SUSPEND, RELEASE,* and many more. There is no reason to list all messages or to discuss all details of these messages here. However, I think it would be useful to present an example of a message because all DSS1 layer 3 messages are defined using the same principles and they contain some parameters that are the same.

Figure 7.5 shows a Call Proceeding message which is used by either a user or a network to communicate the fact that the requested call establishment has been initiated and no more call establishment infor-

* There is a debate in the industry that changes in intensity with time about the correspondence of D-channel layer 3 to OSI layers. Many people believe that a subset of DSS1 layer 3 functionality is really the OSI application layer functionality.

mation will be accepted. It contains a number of parameters, or, in ISDN terms, information elements. Each information element is defined in the DSS1 standard and has its own structure. As one can see, DSS1 messages are defined in tabular form; ASN.1 or any other formal abstract language is not used.

The messages in category (2) are used for supplementary services. These messages are grouped into two classes. The first class contains messages of the separate message approach. This approach is used to define specific messages for specific supplementary services. HOLD and RETRIEVE messages belong in this class. The second class contains messages of the common information element procedure. This procedure uses the *Facility* information element for all supplementary services that do not require synchronization of resources between a user and a network. REGISTER and FACILITY messages, as well as call control messages that contain the Facility information element, belong in the second class.*

In our investigation of a suitable solution to the SCAI transport question, we are looking for messages that are not used for call control, if possible, and contain fields that can be used for user data. Two such solutions are possible: one with call control related messages (category 1) and another with the common information element procedure (category 2).

7.1.1 Common element procedure

The heart of this procedure is the Facility information element, shown in Fig. 7.6.

The use of different fields of the Facility information element is as follows:

- The first octet identifies the type of information element.
- The second octet specifies the length of the information element.
- The five lower bits of the third octet are the protocol profile; it is used to identify the type of protocol that uses the facility-based mechanism (for example, value 10001 is reserved for Remote Operations Protocol).
- Starting with the fourth octet, the information element contains "components," or service-related information.

---

* At this point, some readers may get dizzy from this term *Facility,* losing track of what it is. One needs to remember that DSS1 layer 3 consists of messages, and each message consists of information elements. Somehow the same name was given to a message and to an information element that happened to be used in several messages. The distinction is quite simple as long as one pays attention to the qualifying words following the term *Facility.*

<table>
<tr><th colspan="8">Bits</th><th></th></tr>
<tr><th>8</th><th>7</th><th>6</th><th>5</th><th>4</th><th>3</th><th>2</th><th>1</th><th>Octets</th></tr>
<tr><td>0</td><td colspan="7">Facility
0 0 1 1 1 0 0
Information element identifier</td><td>1</td></tr>
<tr><td colspan="8">Length of facility contents</td><td>2</td></tr>
<tr><td>1
Ext.</td><td colspan="2">0 0
Spare</td><td colspan="5">Protocol Profile</td><td>3</td></tr>
<tr><td colspan="8">Components(s)</td><td>4, etc.</td></tr>
</table>

Figure 7.6 *Facility* information element.

The component field is the place for information generated by the SCAI protocol. It is the kind of user data field that can be used by different applications (services). This field can contain one or more components, and each component is a sequence of data elements. Components are defined based on ROSE. The types of components are:

- Invoke
- Return Result
- Return Error
- Reject

Comparing these components with ROSE APDUs, we find a complete correspondence between the two standards. Each component consists of a tag (a type of component), length, and content. SCAI information (a specific SCAI FE) would be placed in the content portion of one of the components. A simplified format of a Facility information element containing a SCAI FE would look as shown in Fig. 7.7.

Placing SCAI FEs into the content portion of a component seems simple enough. Actually it is not a simple issue and requires some discussion. Two questions need to be answered: (1) What does it mean to use ROSE-like functionality to transport ROSE APDUs? and (2) Should SCAI FEs be individually visible or tagged as *SCAI usage?*

Let's repeat what happens with the PDUs generated by the SCAI functions according to the SCAI standard protocol architecture, also keeping in mind that there are two varieties, with the SCAI Security

direction →

Contents (SCAI FE)	length of component	component type	protocol profile	length of IE	Facility ID

IE -- Information Element

Figure 7.7 *Facility* information element with SCAI FE.

ASE and without it. If the SCAI Security ASE is not used (no security functions are required), the situation may be simpler. The components of the Facility information element perform the functions of ROSE, which means that the SCAI protocol architecture is sustained and is as it appears in Fig. 7.8.

In this case we are using not an independent ROSE, but another mechanism that includes ROSE in it. As long as a manufacturer creates this kind of linkage for all its SCAI solutions, it is workable. However, if a manufacturer wants to create SCAI software that can use different transports, such as DSS1 and X.25, specific software packages have to be created, one for each of these transports. This is an example of how an OSI intention of modular layers that perform only specific functions and therefore do not have to know the details of underlying layers is not fulfilled. The DSS1 protocol, although containing some layer 7 functionality, is supposed to be used by SCAI as a transport only. It should not know or care about the details of the information provided by the upper layers or functions. Yet if ROSE is used, not separately, but in conjunction with other functions, a specific binding with those other functions results.

To avoid the aforementioned problem, if DSS1 is used as a transport, the protocol architecture shown in Fig. 7.9 should be adopted, but that creates another problem.

The problem with this architecture is that it requires that the same ROSE functions be performed twice: once when an original SCAI FE is

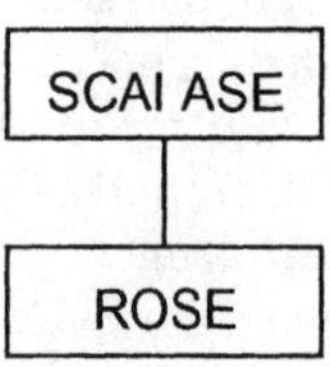

Figure 7.8 SCAI without security (part 1).

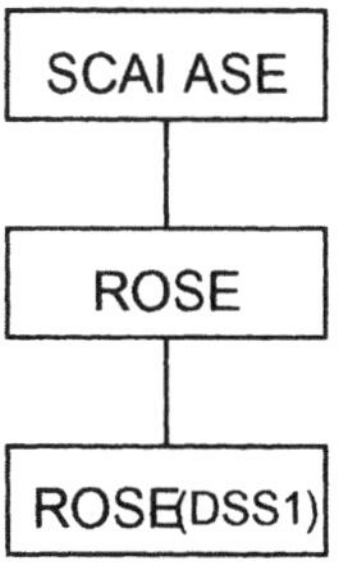

Figure 7.9 SCAI without security (part 2).

being mapped onto ROSE APDUs, and the second time when ROSE APDUs are being mapped onto components of the DSS1 protocol following the common element procedure. This does not seem to be very efficient. The question now—and the answer to this question may help with efficiency—is how to map ROSE APDUs onto DSS1 ROSE components. One way is to map *invokes* to *invokes* and *return results* to *return results*. This idea seems straightforward, but it is exactly this idea that creates inefficiency of duplication. Another way is to map all ROSE APDUs to *Invoke* components in DSS1. So when a ROSE APDU carrying a SCAI Make_Call Request arrives at DSS1 for transport, it is mapped to an invoke component; when a ROSE APDU carrying a SCAI Make_Call Response arrives at DSS1 for transport, it is also mapped onto an invoke component, not a return result component. The two invoke components have nothing in common; they do not "know" that the information they carry pertains to the same invocation of the same function, Make_Call FE. This seems to be a more efficient solution.

Now, let's examine what the situation is if the SCAI Security ASE is also used. The situation changes drastically. According to the SCAI protocol architecture, the security functions affect the entire ROSE APDU. The result is that all information, including protocol elements of ROSE, is enciphered. We do not have the choice of bypassing the independent ROSE and use ROSE as part of DSS1. We also cannot distinguish among different SCAI functions and among different ROSE operations. The only way, in this case, to use DSS1 is to map any secure ROSE APDU, a request or a response, onto an invoke component. When security functions are deployed, the protocol architecture is as in Fig. 7.10.

We can conclude that the best approach in using the Facility information element as a transport for SCAI information is to do the following:

1. Define a new Protocol Profile value specifically for SCAI.
2. Map any SCAI-generated ROSE APDU onto an invoke component.
3. Use only one operation for all SCAI-related transfers—SCAI transfer.

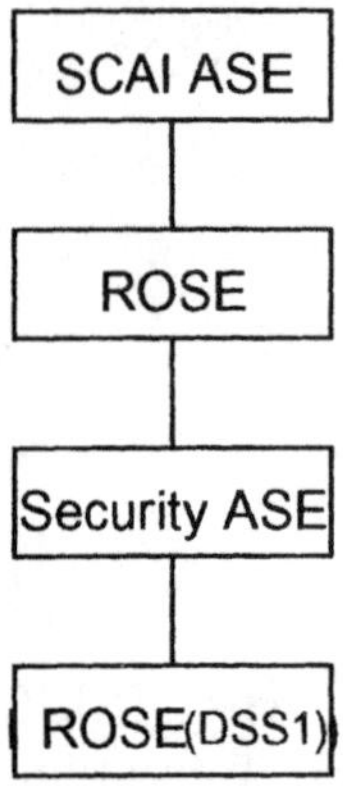

Figure 7.10 SCAI with security.

The protocol profile actually requires some analysis. The value of a protocol profile is used to differentiate among different types of protocols using the procedure. Two choices exist for SCAI. The first choice is to use the existing Remote Operations protocol profile value for SCAI also. This choice would not require any additional standardization efforts and can be applied right now. The disadvantage is that it results in the coupling of SCAI with other users of the common element procedure—for example, ISDN supplementary service—that was never intended. Usually this type of coupling creates problems in the future when (or if) completely different directions are taken by SCAI and ISDN services. The second choice is to reserve a special protocol profile value for SCAI. There are plenty of available values that can be used for SCAI, and technically it is a very easy step. However, this choice does require an agreement by ITU-T and that is not always easy to achieve. Nevertheless, I clearly favor the second choice.

Regardless of the mapping used, three messages can carry Facility information elements with SCAI FEs: FACILITY, REGISTER, and RELEASE COMPLETE. The first two messages are specifically defined for non-call-associated activities, and the last message is required from the procedural point of view. Let's examine the procedure to be used with these messages, but first we need to identify the purpose of each message.

The FACILITY message is used to request or acknowledge a service. When it is used, the Facility information element contained in it carries specific operational information. The format of the FACILITY message is shown in Fig. 7.11.

The REGISTER message is used to assign a new call reference value for a non-call-associated transaction. The reference value is the mechanism by which a user and a network differentiate among all different calls and transactions. It is a mandatory parameter and is always pres-

Information element	Direction	Type	Length
Protocol discriminator	both	M	1
Call reference	both	M	2–
Message type	both	M	1
Facility	both	M	8–
Display	n--->u	O	special note

Figure 7.11 FACILITY message format.

ent. The format of this message is the same as the format of the FACILITY message.

The RELEASE COMPLETE message is used by a user or a network to release a call reference value when an associated call or transaction is cleared. Its format is similar to other messages presented here and contains the Facility information element.

The procedure for using these DSS1 messages to transport SCAI is quite simple. First, a REGISTER message is sent with a new call reference value. This message can also contain SCAI FEs. Next, a series of FACILITY messages is sent to exchange SCAI FEs as required by the application. And last, when communication is complete, the RELEASE COMPLETE message is used to release the call reference. The last message can also contain SCAI FEs. In reality this procedure would consist of a single REGISTER message followed by a large number of FACILITY messages and a single RELEASE COMPLETE message for the duration of required communications between a switch and a computer. Figure 7.12 shows a message flow.

An example of the DSS1 message with SCAI information (FEs) contained in it is shown in Fig. 7.13.

The Facility information element can be successfully used to transport information that would fit into a maximum information element size of 255 octets. If this size is not sufficient, the Extended facility information element can be used. It is coded exactly the same as the Facility information element except for the *length* field (see Fig. 7.14).

The length field can consist of one or more octets. If the length is less than or equal to 127 octets, bit 8 is coded 0 and bits 1–7 encode the length. If the length is more than 127, bit 8 is coded 1, bits 1–7 encode the number of subsequent octets, and subsequent octets encode the length of the information element.

7.1.2 Call Control related transport

Although we identified before that it is better to use a non-call-associated procedure to transport SCAI information, this heading

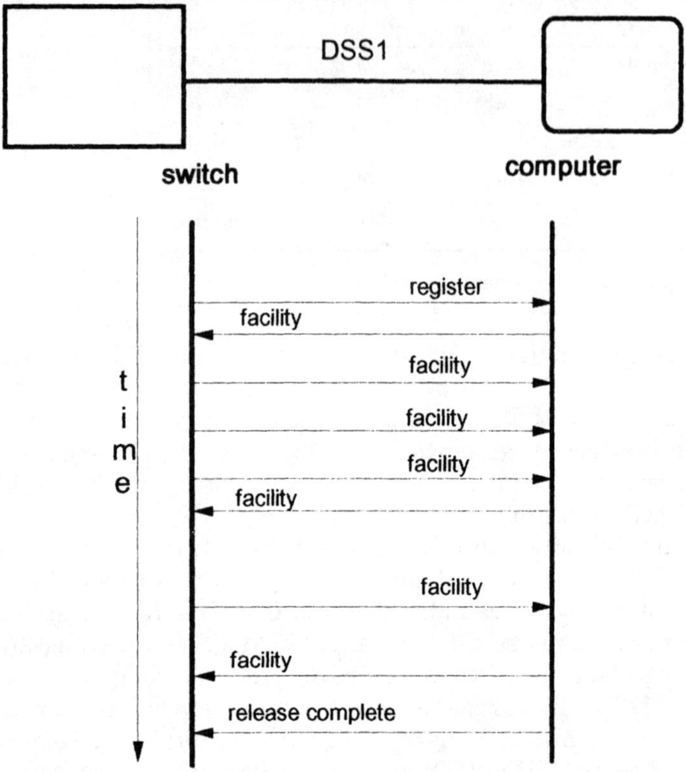

Figure 7.12 DSS1 message flow.

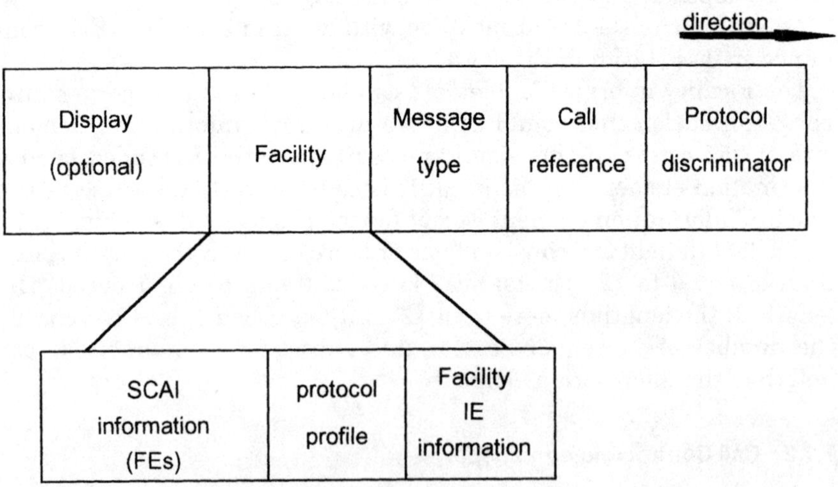

Figure 7.13 FACILITY message with SCAI FEs.

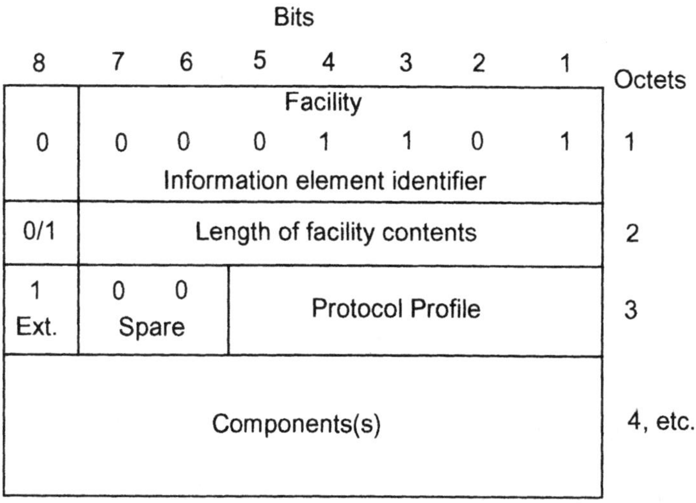

Figure 7.14 Extended facility information element.

seems to contradict this. But it is not the case. As part of call control activity, a special procedure to carry non-call-associated information has been defined. It is called User Signaling Bearer Service (USBS). It is designed specifically for the users' transport needs and can be considered for SCAI.

The USBS procedure relies on several messages for call establishment and clearing, such as SETUP, CALL PROCEEDING, CONNECT, RELEASE. When the call is established, the information is transferred in USER INFORMATION messages, the format of which is shown in Fig. 7.15.

SCAI information can be placed in the User-user information element shown in Fig. 7.16.

SCAI FEs would be placed in the user information field and carried across an interface. The maximum size of this field is 255 octets. If this

Information element	Direction	Type	Length
Protocol discriminator	both	M	1
Call reference	both	M	2–
Message type	both	M	1
More data	both	O	1
User-user	both	M	3–255

Figure 7.15 USER INFORMATION message.

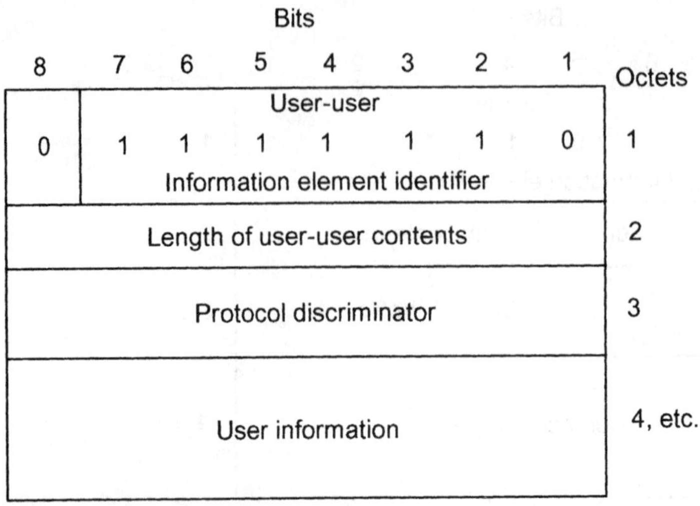

Figure 7.16 User-user information element format.

is not sufficient, more than one User Information message can be used, in which case the More data information element is present to indicate that more messages with information are still to come.

7.2 SS7

Signaling System 7 (SS7) is a standard interswitch signaling protocol. It is standardized in the United States and internationally, and, although there are some differences in details of specific functions, on the whole, the two standards are almost the same. Some terms in the ITU-T and ANSI versions are different and this text uses the ANSI terms. As we discussed before, SS7 is an out-of-band signaling system that uses specific communication channels to carry information among various network entities. Out-of-band network signaling has been used for quite a long time under the general term Common Channel Signaling (CCS). It is "common" because it uses the same channel to convey information regarding many voice/data channels (circuits) used for users' communications. SS7 is the latest CCS standard designed for digital circuits and used over 64 kbps channels. Figure 7.2 shows a signaling channel for an SS7 information exchange and a separate voice/data channel over which an actual users' communication takes place. Although only one voice/data circuit is shown for compactness, there are many circuits that a single SS7 channel "controls."

The SS7 protocol is layered. Like the layers of the DSS1 signaling protocol, not all SS7 layers directly correspond to the OSI layers. The SS7 protocol architecture is shown in Fig. 7.17.

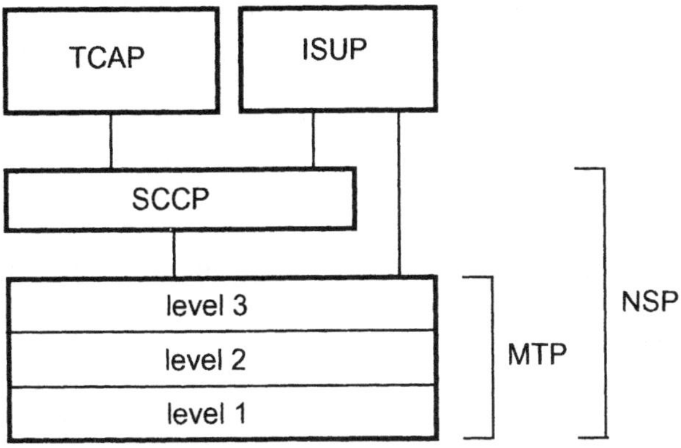

Figure 7.17 SS7 protocol architecture.

The Message Transfer Part (MTP) performs the transfer of SS7 messages between network entities involved with signaling. MTP consists of three levels,* each performing specific functions. Level 1 corresponds to the OSI layer 1 and performs the functions of the physical layer. Level 2 corresponds to the OSI layer 2 and is responsible for message transfer over a single (adjacent) link. Level 3 performs some of the functions of the OSI layer 3 and is responsible for message routing. MTP is not concerned with the message content, its job is to maintain delivery of messages to appropriate signaling nodes in the network. Other SS7 parts are responsible for the meaning (semantics) of messages and these parts are said to belong to level 4 of SS7.

For the purposes of this discussion, it is sufficient to present three parts that belong at level 4: the Signaling Connection Control Part (SCCP), the Transaction Capabilities Application Part (TCAP), and the ISDN User Part (ISUP). There are other standardized SS7 User Parts, such as the Telephone User Part (TUP), but their presentation would not add anything to our discussion.

Although SCCP is being placed at level 4, together with MTP it constitutes the Network Service Part (NSP). NSP is similar in function to the OSI layers 1–3, which are only concerned with the transfer of information. SCCP itself performs the functions of the OSI layer 3. For example, SCCP is used to perform global title translations—identification of a specific number (name) based on a globally visible number (e.g., 800-number translation).

ISUP is the part of SS7 responsible for call control. It generates and receives messages very similar in content to the messages of DSS1

* *Levels* in SS7 is the same concept as *layers* in OSI.

layer 3. As a matter of fact, it is the interworking between ISUP and DSS1 layer 3 that is necessary for the call control information to be conveyed throughout the network. Examples of ISUP messages are *INITIAL ADDRESS MESSAGE* (*IAM*), similar to the SETUP message in DSS1, *RELEASE, ADDRESS COMPLETE,* and so forth. ISUP uses either SCCP, which in turn uses MTP, or it uses MTP directly for message transfer.

TCAP is the part of SS7 created for non-call-associated functions. While ISUP is concerned with the functions of call setup, release, and the like, TCAP is concerned with functions that are not related to a specific call. TCAP is used for relating information that is not part of call processing between network entities, including use by operation and maintenance functions. TCAP uses NSP for message transfer.

Functionally, two SS7 parts can be considered for the purposes of SCAI transport: ISUP and TCAP. ISUP, being a call processing-oriented functional set, is probably not well suited for the job. TCAP, on the other hand, is specifically designed for non-call-associated signaling and can be considered. Let's examine TCAP as a protocol.

From the OSI point of view, TCAP is an application layer protocol. It uses the underlying transport mechanism of SS7 (SCCP) to exchange information between network entities. Although TCAP is not formally defined as an ASE, in fact it is a set of functions that together perform as an ASE. A TCAP message consists of two parts as shown in Fig. 7.18.

Both portions consist of information elements, and each information element is constructed as in Fig. 7.19.

The Transaction Portion of the message contains information on message type (in TCAP's terminology, package type), message length, and transaction information. Transactions are associations that are created for the purposes of a particular communication instance. Each transaction is assigned a reference value (transaction ID) which is used to refer to a particular communication. So, the main purpose of the transaction portion is to identify a transaction.

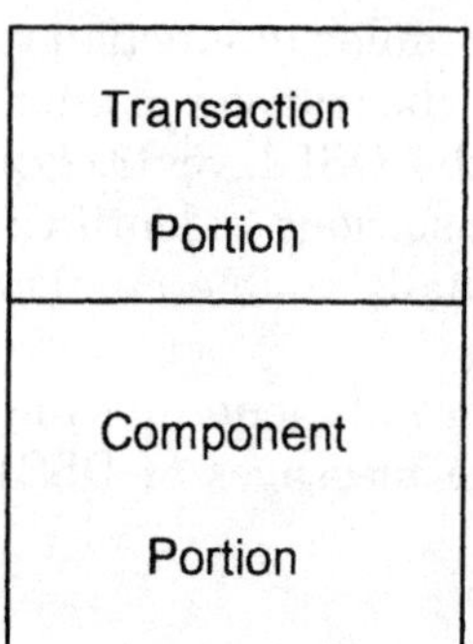

Figure 7.18 TCAP message structure.

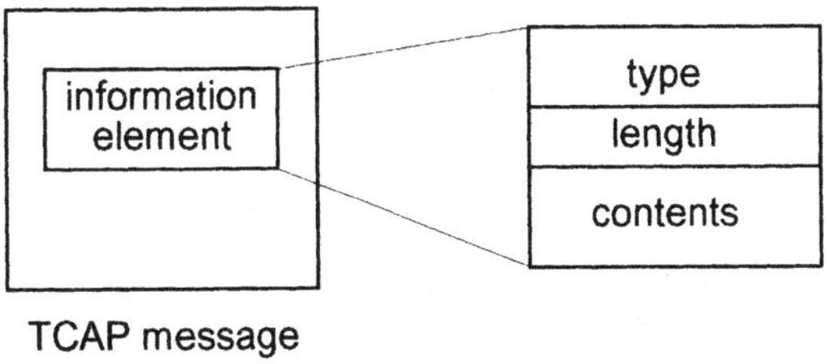

Figure 7.19 Structure of an information element.

The component portion of the message contains information on the actions that are requested from the receiver of the message. It can contain more than one component, each corresponding to a specific request or response. The component portion is designed based on ROSE and contains components of six types*:

- Invoke (Last)
- Return Result (Last)
- Return Error
- Reject
- Invoke (Not Last)
- Return Result (Not Last)

Each component type consists of a number of information elements, of which we are interested in two: Operation Code and Parameters. The Operation Code information element identifies the operation to be performed. The Parameters information element can contain any information. The size of the contents of the Parameters information element is variable and is restricted by the maximum SCCP message size (which is 272 octets) and the number of octets in the overhead of a particular TCAP message. The overhead octets include type identifiers, length of elements, and so on. The available number of octets for information equals (**272 – overhead**). An approximate maximum is 220 octets.

The alternatives of mapping ROSE APDUs with SCAI information onto components are exactly the same as in the case of the DSS1 common element procedure, so we are not going to repeat the discussion.

---

* ITU-T specifies four types of components; ANSI (the U.S. standards body) specifies six types. The *Invoke* and *Return Result* component types in the ANSI standard, but not in the ITU-T standard, can be *last* or *not last,* so there are two additional types. The ITU-T standard uses *Invoke* and *Return Result* components regardless of their position in the component sequence, while the ANSI standard uses the same components with the indication *last* or *not last* component in the sequence. This is an important difference between the international and the U.S. TCAP standards.

It is clear that the Operation Code information element can be used to indicate a SCAI FE, e.g., Make_Call, and the Parameters information element can be used to include FE parameters, given that the SCAI security functions are not used. The approximation of a possible TCAP message with SCAI information would look as in Fig. 7.20.

This message can be "translated" as follows: For the transaction (association) XYZ this is a request number W (invoke ID = W) to perform the Make_Call function using the provided parameters.

Another way of mapping is to create a single *SCAI operation* to indicate that TCAP is used as a transport only. If the SCAI security functions are used, the only alternative is to map all requests and responses to invoke components. This is all the same as in the case of DSS1. However, there is something that the standards committees are beginning to work on that might help: there is work underway on TCAP security functions. If they are standardized, we can bypass SCAI Security ASE as well as ROSE if TCAP is used.

Nevertheless, there are still doubts regarding the use of TCAP as a transport for SCAI. Two issues are responsible for the doubts. The first issue is technical. TCAP contains three kinds of functions: (1) association establishment (transaction portion), (2) operations, and (3) exchange of information between remote entities. While SCAI FEs map easily to TCAP operations, TCAP also performs the functions of ACSE and ROSE. The SCAI protocol architecture defines the use by the SCAI ASE of independent ACSE and ROSE functions, and TCAP combines all of these functions in itself. The simplest way to use TCAP (SS7) as a transport mechanism for SCAI is to bypass ACSE and ROSE and rely solely on TCAP for all information exchange needs. Although it is possible, and we should not be pedantic about changing the protocol architecture since all functions are available anyway, it does detract from the flexibility of using different ASEs in the same computer application environment. It also suffers from the problems of tying a SCAI implementation too closely to a transport mechanism with the result that each transport mechanism would require a separate layer 7 software because it is too closely linked with that transport mechanism.

The second issue is nontechnical, and sometimes a nontechnical issue is more influential. SS7 has been designed as a signaling protocol

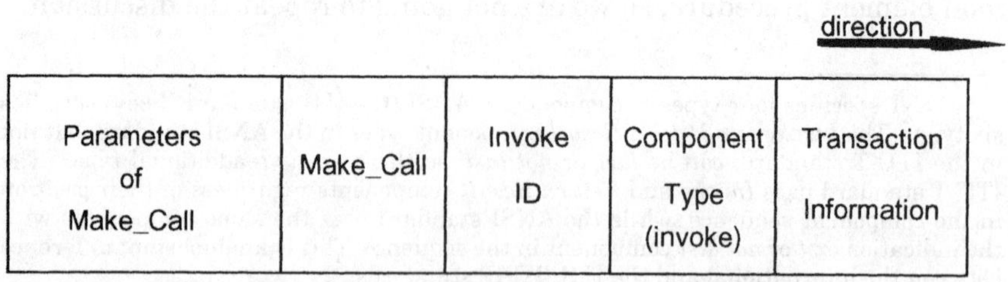

Figure 7.20 TCAP message with SCAI information.

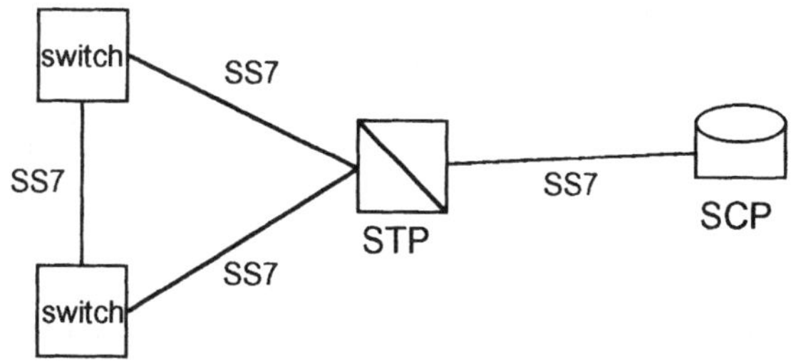

STP -- Signaling Transfer Point;

SCP - Signaling Control Point.

Figure 7.21 SS7 network.

for intranetwork needs; it is used in the networking configurations as shown in Fig. 7.21.

Public telecommunications networks that use SS7 deploy signaling components (lines, databases, etc.) designed for the capacity necessary for their specific signaling needs. Since signaling is a critical part of network performance, public networks are reluctant to allow access to the signaling network unless this is carefully controlled to limit service degradation. SCAI computers, being positioned outside the networks, would gain access to the signaling components of these networks if SS7 is used for SCAI. The only possibility to restrict this access is to create some kind of a front-end SS7 process that would not allow access to the network SS7 components, but this would be highly unlikely.

In the world of private networks, most PBXs are not equipped with SS7 functionality—at least not now. Whether SS7 will be developed on major PBXs or not would be the result of market analysis by their makers and SCAI is only one aspect of this market.

These considerations force us to think of SS7 as theoretically possible, but a practically difficult choice as a SCAI transport mechanism. The only possibility for SS7 to become one is if in a specific environment it is the best economic solution and is already available for use. We cannot neglect this possibility and, therefore, SS7 is presented here as a choice.

7.3 X.25*

X.25 is a widely used standard protocol for accessing packet-switched networks. It is popular not just because it provides the necessary

* There are some differences between the X.25 versions of CCITT standards produced in 1984 and 1988. Here we use the 1988 version (blue book) of X.25.

transport mechanism for data communications, but also because many manufacturers build packet-switching equipment according to X.25 and many networks utilizing this technology already exist. There is a companion standard—X.75—which is used for intranetwork communications. In some sense, the X.25/X.75 pair has a similar relationship as the DSS1/SS7 pair, except that X.75 is an extended X.25 protocol, while DSS1 and SS7 are two different protocols.

X.25 corresponds to the first three layers of the OSI reference model. This is one of the advantages of X.25 because a system implementing OSI can be made to accommodate X.25. Figure 7.22 shows the protocol architecture of X.25.

The physical and frame levels perform the functions of OSI levels 1 and 2; that is, they are responsible for physical and link procedures. The packet level is responsible for packet transfer to a destination. The packet level is of interest in our investigation because it is the level that formats X.25 packets for transmission. The packet level uses the underlying levels to complete packet transfers.

There are two types of packet communications, or data calls, available in X.25: virtual calls and calls using permanent virtual circuits. Virtual Call (VC) service is sometimes called Switched Virtual Circuit (SVC). With this type of call, first the circuit has to be established and then the circuit is used to transport packets of information between a user and a network. Permanent Virtual Circuit (PVC) service uses preestablished connections to exchange packets of information.

Packet level messages can be classified for our purposes into two categories: call setup/clearing and data exchange. For SVC service, both types of packets are used, while for PVC service, only data transfer phase packets are used. Whether SVC or PVC is used, messages used in the data transfer phase should be examined for SCAI.

During the data transfer phase, different types of packets are used for the following reasons:

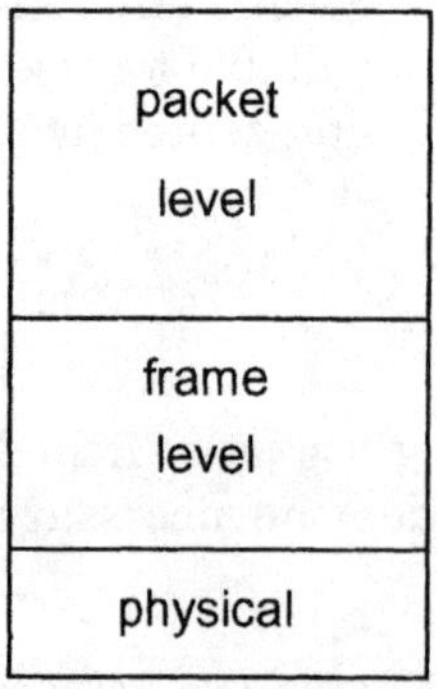

Figure 7.22 X.25 protocol architecture.

- Data transfer
- Interrupt
- Flow control
- Reset
- Restart

The type of packet of interest to us is DATA. It is used specifically to transport data, while all other types of packet are used for various operations, e.g., restart. There are two kinds of DATA packets: (1) for basic service and (2) for extended service. Basic and extended services refer to the number of octets counted in the sequence of sent packets. In case (1) it is modulo 8, and the sequence counter is reset after the eighth packet; in case (2) it is modulo 128, and the sequence counter is reset after the 128th packet. The only difference between DATA packets for the two kinds of service is the size of sequence fields. Therefore, we will discuss only the basic service DATA packet (see Fig. 7.23).

The fields of the DATA packet have the following meaning:

Q-bit—Qualifier bit; it indicates whether user data or control information is contained in the packet.

D-bit—Delivery confirmation bit; it is used to indicate to the network that end-to-end acknowledgment for the packet is requested.

Bits 6 and 5, following the D-bit, when set to 0 1 indicate that packets for modulo 8 service are used.

LCGN—Logical channel group number; together with the Logical channel number it identifies a logical channel assigned for the data call.

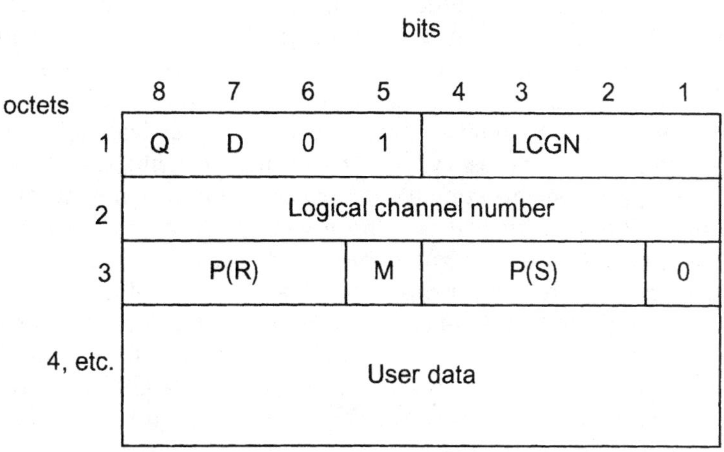

Figure 7.23 DATA packet format.

P(R)—Packets received; it acknowledges the last received packet.

P(S)—Packet sent; it identifies the number of the sent packet.

M-bit—More Data bit; it is used to indicate to the recipient of the packet that more data follow.

Bit 1 of the third octet is always set to 0; it indicates that the packet is DATA; no other packet type has this bit set to 0.

User data—This is the field for user data.

The last field of the DATA packet is of interest to us. This field can be up to 128 octets long and is specified for user data. SCAI messages, or in OSI terms, APDUs, are user data and can be placed into this field. Packets thus formatted are then forwarded to the link layer which adds data link information, and the resulting frames are transmitted over a physical medium. Hence, the X.25 protocol does not require any changes to serve as a transport mechanism for SCAI.

The X.25 protocol is a straightforward alternative as a SCAI transport mechanism. It is defined specifically as a data transport and is well understood in the industry. X.25 is also addressed in ISDN, and ISDN standards specify procedures for X.25 to use on a B- or a D-channel. Therefore, if ISDN is available in a specific situation and X.25 is chosen as the transport mechanism, it can be used as part of the ISDN access interface.

7.4 Frame Relay

The concept of Frame Relay was developed when ISDN standards were still in the works and X.25 was already an accepted form of data communication. X.25 has been designed for use in communications channels of up to 64 kbps. It has many procedures for error detection and correction that are necessary to maintain reliable transfer of data. In the new digital transmission era, the speed and rigorous retransmission facilities of X.25 were seen as inappropriate. Faster transmission speed was found to be necessary and the digital transmission technique did not require such extensive error correction schemes as the X.25 provided. The idea was to develop an ISDN data protocol that would be cheaper to implement and deploy, and, at the same time, can be used on higher-speed links. Frame Relay was born. Before we discuss some details of the frame relaying protocol, it is important to note that initially the idea was that X.25 and Frame Relay would be used for different applications based on the applications' reliability requirements. Applications with stringent reliability requirements would use X.25, and applications with less demanding reliability requirements could use either.

Frame Relay provides SVC and PVC services, just like X.25 protocol. The protocol architecture is very simple because there are only two layers (see Fig. 7.24).

Recommendation Q.922 defines a complete set of the OSI layer 2 functions. If all of Q.922 functions are used, then the Frame Switching service is provided. If only core aspects of Q.922 are used, then the Frame Relay service is provided. The functions in Q.922 in addition to core aspects are sequencing, flow control, and retransmission. Actually, full Q.922 functionality is very similar to the functionality of X.25 level 2. Since Frame Relay, not Frame Switching, is becoming very popular, we will consider only Frame Relay.

The frame structure is shown in Fig. 7.25.

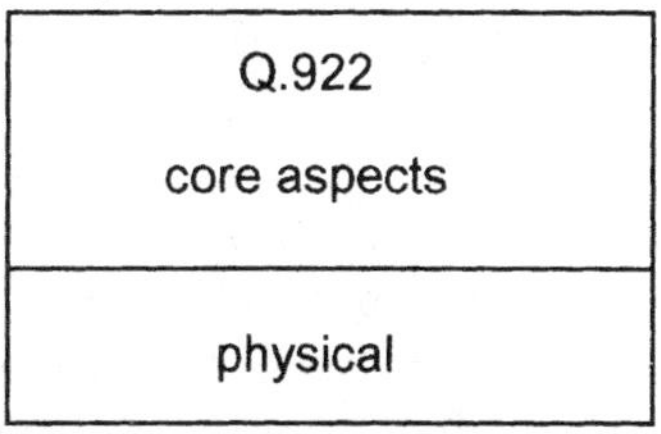

Figure 7.24 Frame Relay protocol architecture.

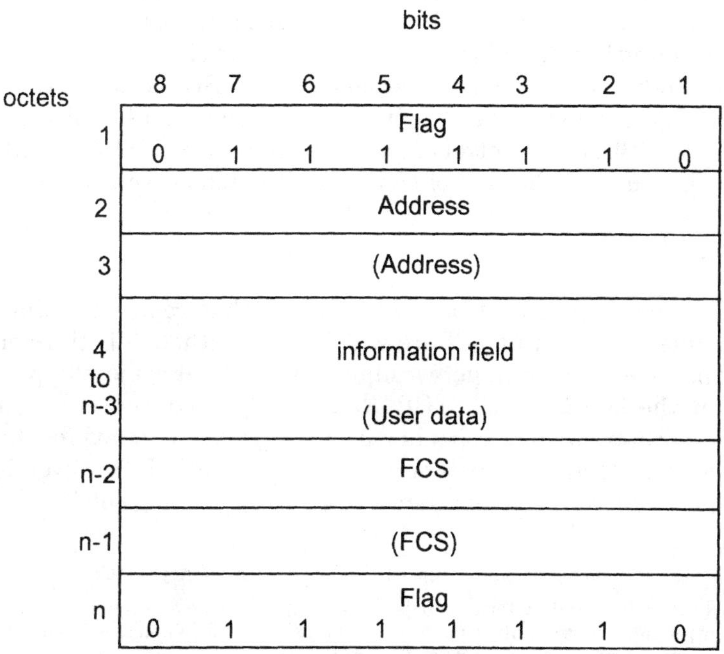

Figure 7.25 Frame format.

The fields of the frame are used as follows:

- Flag—This is a delimiter of a frame.
- Address—This defines a logical connection over which the communication takes place (virtual channel in X.25 terms); this field can be 2, 3, or 4 octets long, depending on the need.
- Information field—This is used to place data to be transmitted; the size of this field is variable and negotiated by the networks involved; the maximum default size is 262 octets.
- FCS—The frame checking sequence is used as a mechanism for checking the correctness of received data.*

Clearly, SCAI FEs can be placed in the information field and transported to the peer system using these frames. The size of the frame's information field can be considerably larger than in the case of X.25 packets, which is an advantage if a SCAI FE is larger than 128 octets.

Frame Relay is a user protocol; that is, it is employed for users' communications. In modeling terms, it is a user plane protocol.† If PVC service is required, then no call establishment is needed and users exchange information frames over a pre-established communication path. If, on the other hand, SVC service is required, then call control procedures specified in ITU-T Recommendation Q.933 are used to establish a communication path, after which the exchange of frames takes place. When a call is finished, Q.933 procedures are used to clear it.

Although Frame Relay has been designed as an ISDN mode of communication, it can be used outside of the ISDN environment also. In such a case, the ISDN call control procedures can be used if SVC service is desired. Of course, in the case of PVC, the situation is even simpler.

7.5 TCP/IP

There is one more obvious transport solution that comes to mind: TCP/IP or Internet. Both names are widely used, although both refer to specific parts of the overall networking system. What is more important is that the system itself, TCP/IP, is widely used and enjoys a greater and greater acceptance among users. This fact accounts for the need to consider TCP/IP as a potential transport for SCAI. However, it is not simple to present a comprehensive view on the usage of TCP/IP

---

* A generic name for such a mechanism is CRC—Cyclic Redundancy Check.

† The ISDN access in modeling terms consists of the control plane and the user plane. DSS1 is a control plane protocol; it has procedures for call control. When a call is established, users communicate over an established path using user plane protocols, e.g., Frame Relay.

by SCAI. First, let's look at the TCP/IP, or Internet, model shown in Fig. 7.26.

As one can see, the Internet model is somewhat similar to the OSI model. It consists of layers performing independent functions. The functionality of the Internet stack of protocols spans from physical to applications requirements. However, the number of layers in OSI and the Internet is different and different protocols are applied at some layers. Both similarities and differences create a situation where it is more difficult to discuss the mutual reliance of two systems than, for example, in the case of analyzing the adaptation of DSS1 to transport SCAI.

The truth is that OSI and TCP/IP are competing networking approaches. In the computing world today TCP/IP has a momentum. OSI, on the other hand, is a more formal, more encompassing, and more general approach. It has the backing and the weight of international standards bodies. Many telecommunications applications and functions are specified using the OSI techniques. The telecommunications industry seems to embrace the OSI more and more. I am not trying to predict a winner in this competition, or even to imply that a winner will emerge. The technology and market forces surrounding this technology are very complex and predictions at this time would not be very credible. Nevertheless, we stated in the beginning of this book that the SCAI standard has been developed using the OSI approach, and the question here is whether TCP/IP can be used as a transport.

The answer can be simple or complex depending on what is meant by "using TCP/IP." Does the usage of only certain layers of the Internet model constitute "TCP/IP usage"? If the answer is *yes,* we can claim that we successfully use TCP/IP when the X.25 protocol is used as a transport. The reason: X.25 is one of the possible *Network* protocols in TCP/IP. Another possibility at the Network layer is Ethernet, a LAN protocol which is also widely used. If the answer is *no,* we need to see if higher-layer protocols are adaptable to SCAI needs. These protocols

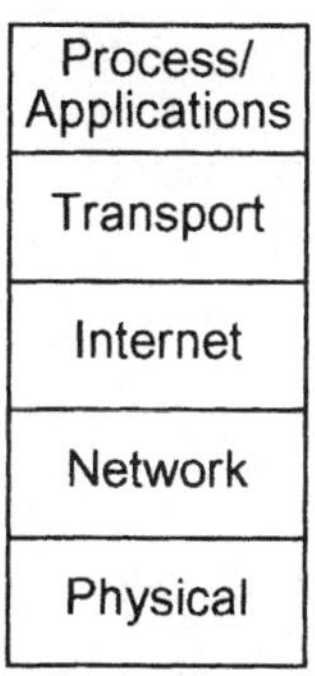

Figure 7.26 Internet model.

are TCP at the Transport layer and IP at the Internet layer. The Transport layer performs the functions of the OSI transport layer and some functions of the OSI Session layer. The TCP protocol, therefore, handles data consistency functions. The Internet layer performs some of the functions of the OSI Network layer. The IP protocol handles addressing and routing.*

The Internet model also contains the Process/Applications layer which performs the functions of the OSI Applications and Presentation layers, and parts of the Session layer. One can claim that to use TCP/IP properly, the Process/Applications layer must also be used. But we have defined the SCAI application layer functions as OSI elements. How these OSI elements can be merged with the Internet higher layer requires detailed analysis. Or does it? Are we making too much of the formalities of two approaches, OSI and TCP/IP? Can't we just simply use the protocols that make sense in a particular environment?

The practical answer at first glance should be *yes*. But one has to be careful, because very simple and practical decisions made today sometimes are not such good decisions, and hindsight lets us understand it only too late. The problem is that technology and services evolve and we attempt to design our systems in a way that will allow us maximum flexibility in the future with the least effort. Unfortunately, merging pieces of different systems or approaches together may lead us to the wrong solution.

Comparison of OSI and TCP/IP from various points of view would take an entire book, and this would probably be a timely and useful book. I am not trying to compare them here or to suggest definite conclusions regarding the TCP/IP usefulness for SCAI, just to raise some issues and focus attention on the TCP/IP as a definite possibility in a real-life situation.

---

* TCP and IP are the key protocols in the Internet suite of protocols and gave the entire system the name. These protocols can be viewed as the distinguishing characteristics of the Internet that make it unique.

Chapter

8

How to Apply the SCAI Standard

When any technology is chosen for a specific job there are still issues to be clarified with respect to requirements and arrangements particular for an environment in which the technology is to operate. This is no different for SCAI. When the SCAI standard is selected to support specific applications, certain issues have to be analyzed. Although many of these issues are very specific to the organization that plans to install the SCAI equipment (e.g., specifics of the applications), some issues are generic enough and can be discussed here.

If the applications under consideration are of the types that the SCAI standard currently supports, such as call management, then the process of analyzing details of the specific requirements can begin. If the application description is not a textbook SCAI application and there is a need to analyze SCAI as a potential protocol, certain steps can be taken to do that. The latter situation is covered in Sec. 8.1, while we begin to identify the process of adapting SCAI to local requirements, which is the same in all cases after the decision on SCAI selection has been made. The following are the steps of this process.

1. *An application context has to be selected.* The SCAI standard identifies two application contexts with distinct object identifiers: one containing only SCAI ASE functionality and another containing SCAI and Security ASEs functionality. The choice between the two standard application contexts is based on the security needs in a particular environment. If, for instance, encrypting equipment cannot be used, there is no need for the application context with the SCAI Security ASE.

The selection of a standardized application context guarantees interworking among systems made by different manufacturers that use the same application context. Does it mean then that all of the SCAI FEs have to be exchanged in any application using one of the SCAI standardized application contexts? The answer is *no.* Only FEs that are needed for a specific application can be used. The fact that the application context contains certain functionality does not imply that all functions must be used. If the application does not ever need to issue a Make_Call FE, it is not contradictory to the use of the application context that contains this FE. In addition, any event report, such as Call_Delivered, Agent_Logged_On, Call_Held, can be included or excluded in the monitor setup, so a particular event report may never be issued if it is not required by an application.

Selection of the standardized SCAI application context is possible when the required functions are only those defined by the standard. If for any reason a manufacturer has to define privately required and, therefore, not standard functions, a nonstandard application context must be used. There exist two alternatives for creating a manufacturer-specific or a user-organization-specific application context for SCAI. One alternative is to use the SCAI Object ID and standardize an object ID value for a specific organization underneath it. The resulting object identifier path looks like Fig. 8.1.

The SCAI standard contains a number of modules (see Secs. 5.6 and 5.7) with object IDs derived from the SCAI object ID, e.g., the SCAI Application Context One has an object ID value (1 2 840 10021 8). Similarly, a specific application context for an organization (ORG) is assigned an object ID value (1 2 840 10021 XXX). The T1S1 subcommittee would be responsible for the assignment of the XXX, and the new object ID value would be unique in the universe of SCAI standard users. This approach would allow any organization that desires its own application context to be publicly recognized to maintain a unique identity for SCAI. The details of the XXX application context are spec-

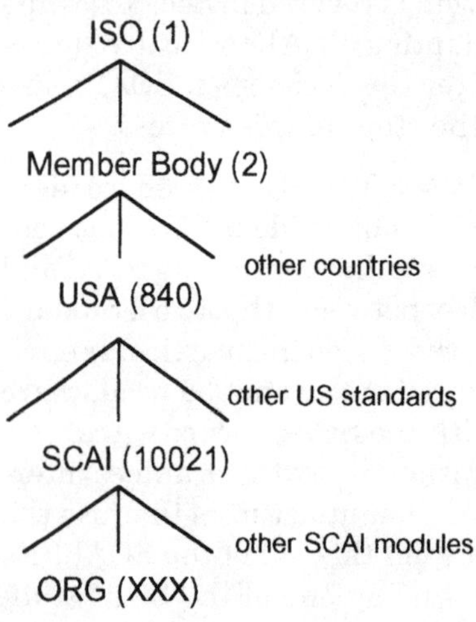

Figure 8.1 Organizational SCAI object ID (part 1).

ified by the organization that "owns" it and any other organization willing to use the same application context would have to learn the details from the creator of the XXX. It is a similar situation as with the SCAI object ID which was assigned to the SCAI standard by ANSI, and the SCAI standard identifies its details.

The other alternative to acquire a unique object ID is to follow an international object identifier tree as we discussed in Sec. 4.2.1. One of the branches of the ISO node is called *identified organization.* If an organization registers under this node, its object IDs are uniquely identified worldwide. If a manufacturer desires to create the SCAI object ID unique to this organization, it can use this alternative and the result would be as shown in Fig. 8.2.

The *Goodcompany,* according to Fig. 8.2, "owns" an object ID specifically for its SCAI implementation. Its value is:

```
{iso identified-organization(3) goodcompany(ZX) scai(W)}
```

This object ID is unique worldwide (ZX and W would be integers in a real situation); if the organization wants to interwork with other manufacturers' equipment, the specifics of this object ID (its functions and rules) have to be shared with those companies.

In either of the cases discussed so far the creation of a specific organizational SCAI object ID does not imply that the SCAI standard is not followed. Theoretically it is possible to create any object ID, but practi-

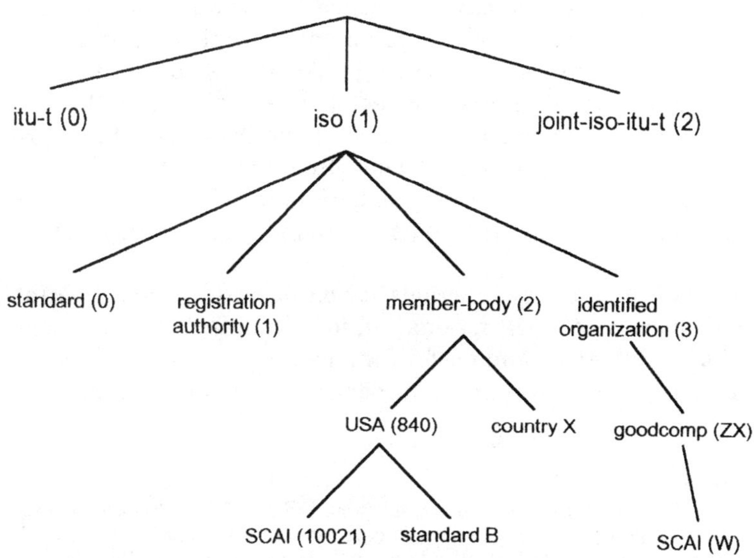

Figure 8.2 Organizational SCAI object ID (part 2).

cally there is no reason for it. What I am describing should be used in a situation when a company, either a manufacturer or a user, wants to add functionality to the SCAI standard without going through the standard process.* The result would be an implementation of the SCAI standard with some functional additions.

Now, when an application context is selected, the optional items from the SCAI standard have to be identified and selected.

2. *All options have to be selected.* There are a number of options that are possible and they have to be geared to particular applications and a particular environment. First, a monitoring option should be identified, that is, whether monitoring would be preset by subscription or set dynamically after the system startup. Although SCAI monitoring functions are flexible and can be used dynamically to change monitors and filters, it is necessary to identify how the startup will work.

At this point we selected the application context and know which SCAI functions will be used (let's assume that the SCAI Application Context Two with the Security ASE is selected). Most functions (FEs) have optional parameters and the decision must be made which parameters are needed for a specific application and which are not. Sometimes, parameters are optional because it is not known whether the system would have the required information. In these cases, nothing can be done; the parameter would be filled if information is available and vice versa. In other instances, the information may be available but not needed by an application, or, although the parameter is optional, the application absolutely requires their presence. In the former case, the agreement may be reached between the manufacturing parties involved with the SCAI equipment not to fill certain parameters even if the information is available. Another solution is just to allow the application to ignore unneeded information, but the first alternative is better because less data would be transmitted and processed. In the latter case, the equipment's functionality has to be carefully checked against the potential problems of missing those required optional parameters. If any problems are found, adjustments have to be made.

In the case of the security functionality performed by the SS_Transfer FE, most parameters are optional. If, for example, there is no need in a specific private environment for functionality responsible for connection integrity and connection confidentiality, the optional parame-

* It is predictable that various manufacturers would like to differentiate their products from other competing products by introducing new features. They would not like these features to be standardized because then anybody can implement them and the market advantage disappears.

ters *confidentiality algorithm, sequence number,* and *time stamp* should not be used.

3. When functional aspects of SCAI have been analyzed, the next step is to *select a transport mechanism for SCAI messages.* The required capacity of the interface has to be determined based on the projected number of service calls, the type of service calls, and the sizes of messages involved in the service calls. The number of calls can be projected as a function of the number of agents (or operators, or other users) to be employed. The type of service calls is known; it is derived from the application itself. If it is a customer service center application, for example, the type of messages are event reports, requests for holding and transferring, and so on. It is now easy to calculate an average message size per call and, further, an average number of messages per call. These calculations would give us an average number of octets to be transmitted per hour. Based on this figure a required channel bandwidth can be calculated.

Each potential transport mechanism is a protocol which also has an overhead. This overhead has to be included in the process of bandwidth calculation. For example, in the case of X.25, the overhead is a few octets at a packet level and a few at a frame level. In the case of SS7, there is overhead associated with TCAP, SCCP, and MTP messages. The overhead may not be critical if the overall requirement for transmission capacity is not high, but it can make a difference if the required capacity approaches the bandwidth of the available channel.

And this takes us to the next item in our process of selecting a transport mechanism: What is available? It may be that in a particular environment a particular mechanism is already available and used for other needs. Its capacity can then be increased or it may have extra capacity. When existing systems are used, the expenditures for training and other administrative efforts are smaller. As we discussed before, a variety of transport mechanisms can be used for SCAI. The one that is available already is probably the first choice, provided it has sufficient capacity.

What if ISDN is available in a particular environment? What kind of ISDN transport should be used? We have identified four ISDN possibilities: the USBS procedure, the common element procedure, X.25, and Frame Relay. All are viable alternatives, each with different overhead and different advantages and disadvantages. Practical experience is required to learn more about the use of these different alternatives. However, it is probably possible to state that the system that has been designed for a specific purpose would be more appropriate than the system that has to be adapted. Although I do think that this assumption is not sufficient for a final selection decision, if the decision has to be made without the benefit of practical experience, I would select either Frame Relay or X.25.

8.1 Is SCAI for Me?

This is not a general question, but a question related to applications that are not explicitly supported by the SCAI standard. We know that SCAI can support call management and other applications we discussed. The variety of applications is great and most of their names are not descriptive enough to really understand what the application is doing, for example, message desk. When an application is discussed and there is a need to identify an existing communication protocol that can support it, what are the criteria by which we decide whether the SCAI standard should be considered?

I was asked the question "Is SCAI for me?" many times. Most times the question was asked because the words "switch" and "computer" are generic enough to lead people to immediately consider SCAI. Each time I had to evaluate the question using the following criteria.

First, the functionality of the targeted application has to be understood. The current SCAI standard focuses on coordination of processes in telecommunications and data processing environments. It is easy to check if the application requires any coordination. If it does, one criterion is satisfied. An additional test is to check if the application can benefit by using third-party call control functions. If it does not, it is not a disqualification for SCAI (if the first criterion has been satisfied), but if it does, the case for SCAI is stronger.

Second, the service configuration has to be analyzed. This is more important to the potential public network applications because of the variety of possible configurations (inside networks vs. outside networks). If the considered application is to be located outside the network from the regulatory point of view,* the second criterion is satisfied. This is not a very strong criterion because it is possible to deploy SCAI within a public network. However, if the computer hosting an application is outside a network, it makes the case for SCAI stronger.

Third, the need for security has to be analyzed. If the planned interface requires security services, SCAI is a good candidate because of its built-in security functions.

Fourth, the desired protocol architecture should be analyzed. If compatibility with OSI and flexible use of various protocol components is desired, SCAI is a good candidate.

And, fifth, the potential transport requirements should be analyzed. If flexibility of transports is required because of the projected evolution of the existing transport mechanism, again, SCAI is a good candidate.

---

* This is a question of ownership. Regardless of where equipment is located, if it is a part of a public network (belongs to a public network), it is inside the network; if it does not belong to a public network, it is outside the network (see Sec. 5.3.2).

Five criteria are identified for the decision-making process of selecting SCAI as the protocol to support a considered application. The first criterion, though, I consider the most important. If it is not satisfied, the potential of the current SCAI standard for an application support is minimal, if not nonexistent. On the other hand, if any of the other four criteria are not satisfied, SCAI can still be selected. However, the more of these criteria that are satisfied, the stronger the case is for the SCAI standard to be considered.

The five steps in deciding whether SCAI is applicable to a particular application do not include checking the SCAI functionality and comparing it with required functions. The preceding algorithm can be used in deciding the general question of SCAI applicability. If, after going through these five steps, the decision is to use SCAI and some functions are not available, three options exist:

1. Do not use SCAI.
2. Add necessary functions (or modify existing functions) and use a private solution.
3. Attempt to standardize additional functions in T1S1.

The option must be selected based on the specific circumstances which include a particular functional discrepancy between SCAI and required functionality, timing of desired deployment, available resources, and other similar considerations.

Chapter

9

The Question of IVR

So far we discussed an integration of computing and telecommunications environments without mentioning a device that is treated as a potential integration key by many. The device is the Interactive Voice Response unit (IVR).* Although IVR is an extremely important part of the reality of computer/switch integration and will continue to be important, it is not the key element of it, but just another component.

An IVR unit is a computerized device capable of maintaining an interactive dialog between itself and a caller. An IVR accepts caller's input usually in the form of dialed digits. Based on the received digits, it instructs the caller about alternative actions or asks for more information. This is achieved through a kind of announcement system that uses prerecorded speech. The types of possible input and possible output will definitely change in time with technological progress in voice recognition and voice synthesis. The new technology should spread the use of IVRs, but the reasons for their use would probably not drastically change.

Three types of IVR use can be identified. The types are distinguished by the goal of a specific human/IVR interaction. These types can be named as follows:

1. Preprocessor
2. Announcer
3. Information collector

The first type, the preprocessor, is probably the most widely used type. The preprocessor IVR is connected to a caller before the caller

* Sometimes it is also referenced as VRU—Voice Response Unit.

gains access to an operator/agent/etc. The IVR, in this case, performs the duties of a human operator and allows an operator to concentrate on real questions and problems instead of spending time on collecting information or providing information that can be as easily performed by an automated machine. The IVR asks a caller to input certain information (e.g., identification number), presents a caller with alternatives (e.g., gaining access to department A or department B), or provides a caller with prerecorded information that otherwise would have to be provided by an operator. Two possibilities exist when this type of IVR is used: (1) a caller is satisfied with the results of the interaction and hangs up and (2) a caller still needs access to an operator. In case (1), an IVR completely substitutes for a human being and resolves the question which triggered the call. In case (2), an IVR collects information that an operator can use when a caller and an operator are finally connected. In both cases, an IVR saves resources because in a large operation fewer operators are required if IVRs are used.

The second type of IVR, the announcer, is used when the purpose of its connectivity to a client is to provide the client with information. For example, in our discussion of the possible industrial control application, the use of an announcement system was mentioned as a possibility for conveying the problem to a responsible person.

The third type, the information collector, is used specifically for purposes of collecting information only. It usually allows an escape to a live operator as part of its operation, but it really is designed for collecting information and is used in situations when callers need only leave appropriate information to receive service. For example, government agencies use this type of IVR to allow citizens to request some government publications (tax-related, and the like). This type has similarities with a preprocessor primarily because it has to be capable of handling exceptions. For example, since most IVRs ask for input in the form of DTMF* digits, if a caller happens to have a rotary telephone, he or she must be connected to a live operator who can collect information or answer questions.

Why are these devices considered to be integration devices? One answer is that they allow access to computers from a telephone network. IVRs are computers capable of processing a lot of information; in addition, they can be interconnected with applications hosts that perform more functions. However, thinking of IVRs as the sole devices needed for integration would be incorrect. IVRs are necessary and

---

* Dual Tone Multi Frequency (DTMF) is a technique of sending digits as pairs of tones. Each digit has a different pair of tones assigned to it. Hence, it is easy to distinguish the digits and the operation is much faster than with rotary dialing. All modern phones use DTMF.

serve an important function, but they are not sufficient for the integration of computing and telecommunications environments. The link merging the two environments must be more general, more flexible, and more functional to support the required integration. So, how can IVRs be used with SCAI?

At the present time there are no provisions in the SCAI standard for IVR connectivity. This is the case, not because it was considered unnecessary, but because of lack of time. Time is mentioned again and again in this book as the reason for the lack of certain capabilities. Unfortunately, it is a fact of life in standards and will remain so. The timing question is always related to the question of priorities. The identification of appropriate priorities is a critical task. Since it was known that IVRs are used in the CTI industry in conjunction with ACDs and ACDs were not the subject of standardization, the decision was to postpone the discussion of other IVR/SCAI capabilities until later. The result is that in the SCAI environment today an IVR is associated with an ACD and usually works as an announcer. An example of such a use is the announcement: "*You are number five in the queue; projected wait time is two minutes.*"

An interesting question is how an IVR would be interconnected with the SCAI entities and what would be the positive result in functional terms. An obvious connectivity scenario is shown in Fig. 9.1.

An agent is a representation of a SCAI user (see Chap. 4). An IVR unit is connected to both a switch and a computer. Connectivity to a switch is required because an IVR has to be connected to a voice path to receive or provide information. Some processing is done by an IVR and then the information is passed to a computer; hence, the connec-

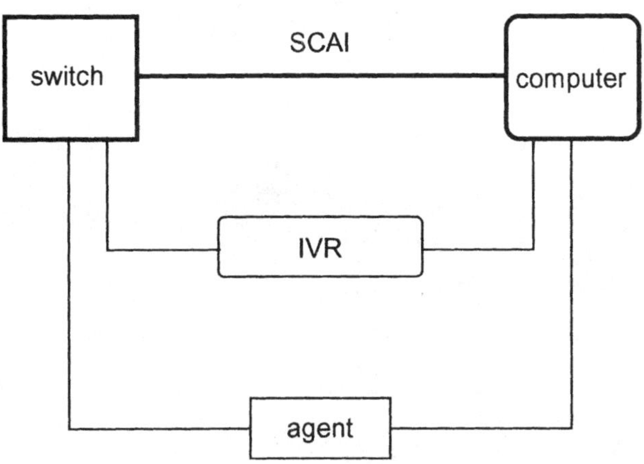

Figure 9.1 SCAI configuration with IVR.

tivity between an IVR and a computer. The functions that an IVR in the shown configuration performs would depend on the application it is used for. One important function could be the acquisition of a customer's identification number to be used by a computer application to locate computer files associated with the customer in the absence of an automatic form of caller's identification. Since this issue is not currently covered in the SCAI standard, it is an open question how to perform such a function. The critical part of this function is the synchronization of the call IDs with caller's IDs at the time of IVR involvement, which is before SCAI functions are used and, therefore, before messages are exchanged between a switch and a computer.

Chapter

10

Future Developments

What can the future of SCAI as a solution to CTI be?

We are discussing a very complex subject—the seamless merging of what we call two environments, telecommunications and data processing, and they are not static. There are new developments in hardware and software technologies, changing regulations, shifting markets. All these factors have to be taken into account while attempts are made to move toward a single information environment. Demand by the public of a certain level of services and changing of public expectations are also part of the equation. The degree of users' demand for new applications is impacted not only by the state of technology, but also by the state of the economy.

Any forward-looking planning requires the acceptance of some assumptions based on which the technology evolution is developed. We should assume that demand for new and more sophisticated services will grow, that those new services will have to be supported by a seamless single environment, that tools for using a single information environment will be required. These assumptions lead us to the issue of SCAI requirements in support of new services. We should approach this issue from two directions: (1) enhancements of the SCAI standard to support short-term needs and (2) additions to the SCAI standard to support long-term needs.

10.1 Short-term Needs

Short-term needs are needs that are obvious right now, but the current standard does not fully address them. These needs are derived from the type of applications that we discussed in Chap. 6. Some of these needs are already mentioned in the appropriate chapters as items "in the works." A complete list of the potential enhancements for the SCAI standard is as follows:

1. Conferencing of *n* parties. The existing SCAI standard allows conferencing of only three parties. Although it is sufficient in many cases, a more general conferencing of any number of parties should be added.
2. Call forwarding. It is a widely used feature. The SCAI standard contains an event report indicating that a call was forwarded (diverted). The call forwarding feature itself should be added to the standard. Call forwarding can be defined in several "flavors": forward always, forward when busy, forward when the calling party is a specific number, and so on. It is probably useful to define as many variations of call forwarding as possible to permit all kinds of service alternatives.
3. Visibility of trunks. Currently, only events occurring at the line side of a switch may be visible and reported. What is taking place at the devices that are participating in calls, but connected to a SCAI switch over trunks, is not completely known and has to be assumed. Figure 10.1 shows an example of such situation.

   Since a SCAI interface connects a host to switch A, line A can be monitored (it is equivalent to saying that a device at the end of line A can be monitored). Line B, connected to switch B, is participating in a call with line A, but is associated with switch A through a trunk. Since the trunk is not currently a SCAI object, events associated with any devices that are reached over a trunk cannot be seen. Adding a trunk object would alleviate this problem. In addition, adding a trunk object would allow the function of selecting a specific trunk group for routing to be defined.

   Adding a trunk object to the SCAI standard does not guarantee the visibility of remote devices. It can only be achieved if a signaling system controlling a trunk is capable of transporting the necessary information regarding the behavior of a remote terminal. If such a

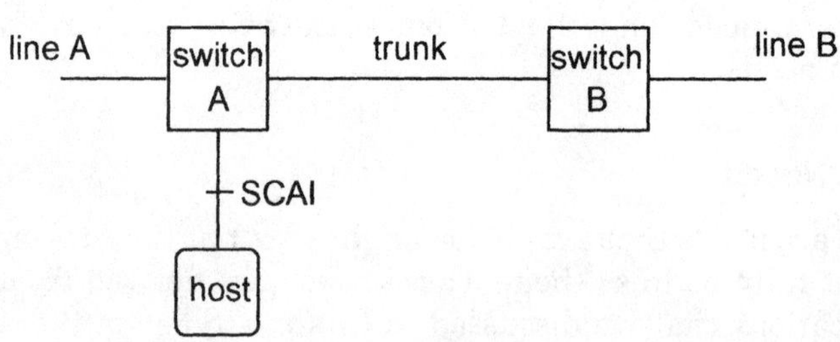

Figure 10.1 Trunks and lines.

signaling system is available (SS7 would be able to do the job), a work at home arrangement is possible. Some of the agents in telemarketing and customer service applications could stay at home and work as part of a department that is located miles away. Since work at home is becoming ever more popular and progressive because it saves time and benefits the environment, this capability may be very desirable.

4. Management. Currently, there are no management functions in the SCAI standard. Although SCAI entities (switches, computers) can be managed by network management systems from the network point of view, local SCAI management is also required. It would provide provisioning and statistics gathering functions, which can interwork with appropriate network management systems.
5. IVRs. As we discussed in Chap. 9, IVRs are not included in the current SCAI standard and should be added to the SCAI architecture and functionality.
6. Determinism of call delivery. Currently, there is no function in the SCAI standard that is invoked to perform a totally deterministic simultaneous delivery of voice calls and computer information screens (data calls) to agents. Although it has not been viewed as a critical function, its existence may be very useful, especially if a network of computers is involved in the decision-making process.
7. Networking. Currently, the SCAI standard specifies functionality sufficient only for point-to-point configuration. Multiple switches and computers may have to be interconnected and new functions have to be defined to handle this more complex arrangement.
8. New FEs (messages). These have to be defined based on the applications that are not currently covered by the SCAI standard. It is impossible to imagine all applications that may require SCAI support, but at least the applications mentioned in Chap. 6 that are currently not supported by the SCAI standard should be considered.

10.2 Long-term Needs

Long-term needs may have to be satisfied in a not so distant future. I have chosen to make a distinction between two categories of needs to separate the items that are known to be missing in the current SCAI standard, although they have been mentioned before, from the items that would signify a fresh view at the integration process.

Several possible directions exist for the evolution of the CTI technology and the SCAI standard. Predicting which direction would dominate is as difficult in this case as in other cases of technological development. Remember when VCRs became popular? In the beginning, two formats were introduced: VHS and Beta. All technical signs were pointing to the Beta format as superior in quality, but the market decided otherwise. The result of the VCR market fight was partially influenced by the widespread appearance of video stores carrying a large number of VHS movies. We all know where Beta VCRs are now. Although this example may sound like an oversimplification, the market's influence outside of technology cannot be underestimated.

Today's CTI solutions and the standard are geared toward ACD applications and other applications supported by large organizations for what I can call "global" purposes. Think of the examples of industrial control and emergency applications in Chap. 6. These are global applications in the sense of serving an enterprise. One possible direction for evolution is to create an information environment which would allow individual users to freely access information in computers from their telephone devices, or to access telecommunications services from their individual personal computers (PCs). We can imagine SCAI linking a telecommunications network with a computer network which is built as a LAN. Each computer on the LAN is a PC serving a person (user). In a sense, we can state that a SCAI application in this configuration is distributed among many computers. This is different from the existing notion of SCAI as a link between two environments, each represented by a single entity: a switch and a computer. Part of this notion is an understanding that a SCAI-linked computer can be networked with other computers for the purposes of sharing or delivering information. But it is not the same as saying that all PCs have equal access to the SCAI link. Figure 10.2 shows a possible configuration of PCs organized into a LAN and linked by SCAI to a telecommunications network.

In this configuration, each user has a telephone and a PC, and all PCs are connected into a LAN. The LAN is connected to a switch with a SCAI link. The SCAI link connects a switch to a data processing environment, but more than one computer (PC) communicates with a switch. How the communication over the SCAI link is accomplished should be a matter of implementation. Some implementations may allow only one PC at a time to communicate with a switch, using the algorithms of a specific LAN. The PCs would either compete for permission to communicate or take turns. Other implementations may allow many PCs to communicate with a switch simultaneously. In this case, a switch would have to be logically partitioned into a number of applications communicating with specific PCs. The SCAI link would have to be engineered based on the need for multiple simultaneous communications.

Using this configuration would allow owners of PC/telephone pairs to access telephony services from their PCs, and the PCs would be used

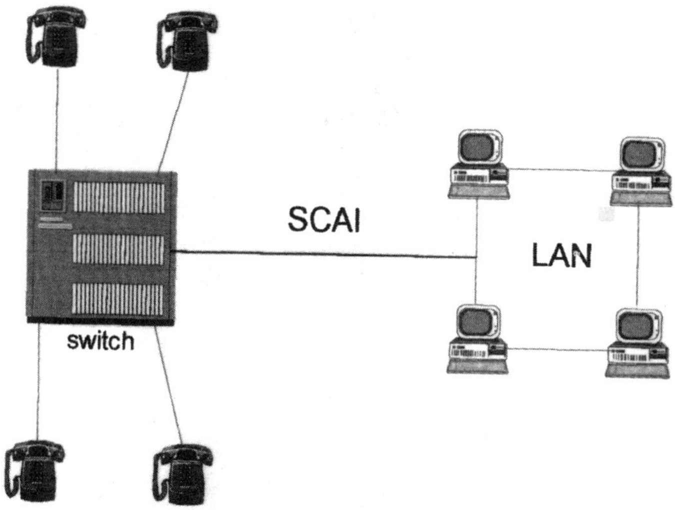

Figure 10.2 LAN/switch interconnection.

both as computers running SCAI applications and as computer terminals. Because a PC is used by the owner for different purposes, not just SCAI, it would be reasonable to predict that the same PC would be equipped with multimedia capabilities, for example, voice and e-mail messages could be converted from one format to another, depending on the particular needs. Merging the multimedia computing capabilities of PCs with the networking abilities of telecommunications networks would create an environment supporting message interchange among many users located anywhere in the world, provided the networks of the world are built to the same standards. Messages could be encoded and decoded using local multimedia capabilities and shared in any form with anybody thus equipped.

A variation of the same type of arrangement is a single computer connected with a switch through a SCAI link and representing a number of PCs as a telephony server. The difference in this case is that all PCs are controlled, for SCAI purposes, by a single server, and this server acts as a LAN gateway from the SCAI link's point view (see Fig. 10.3).

Another possible direction that the CTI evolution may take is the creation of a single information environment through a different type of SCAI link. So far we have discussed a SCAI link being an interface directly connecting a number of entities (switches and computers). Even in Figs. 10.2 and 10.3 the SCAI link is a single link between two environments that can be constructed in many different ways. Now think of a SCAI being not a link, but a LAN. All entities would be connected in a fashion that allows their direct participation.

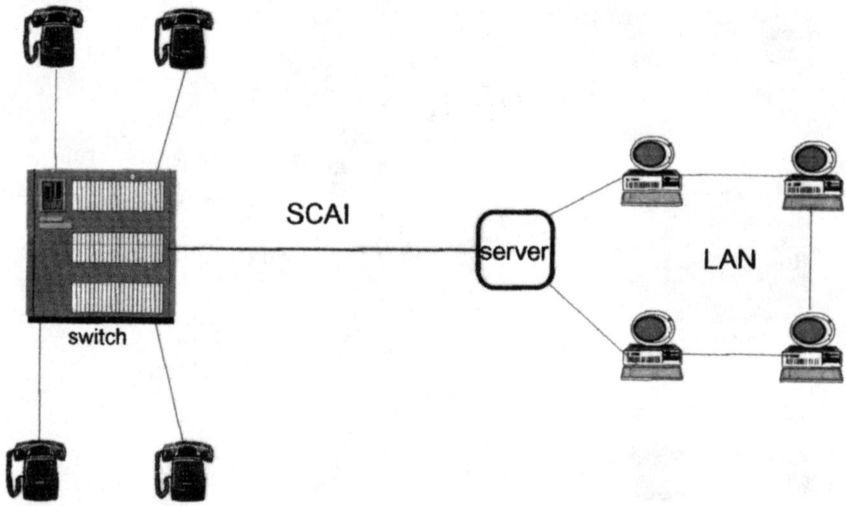

Figure 10.3 LAN/switch interconnection with a server.

In Fig. 10.4, SCAI is actually a LAN: a switch, a PC server, a fax server, and a video server are interconnected into a LAN. In addition to already familiar functions performed by a switch and a PC server, fax and video functions are supported by the respective servers. The fax server is responsible for sending and receiving faxes, and the video server is responsible for a "great new video service." Other types of servers can be added, but the point is that all of these entities are equal

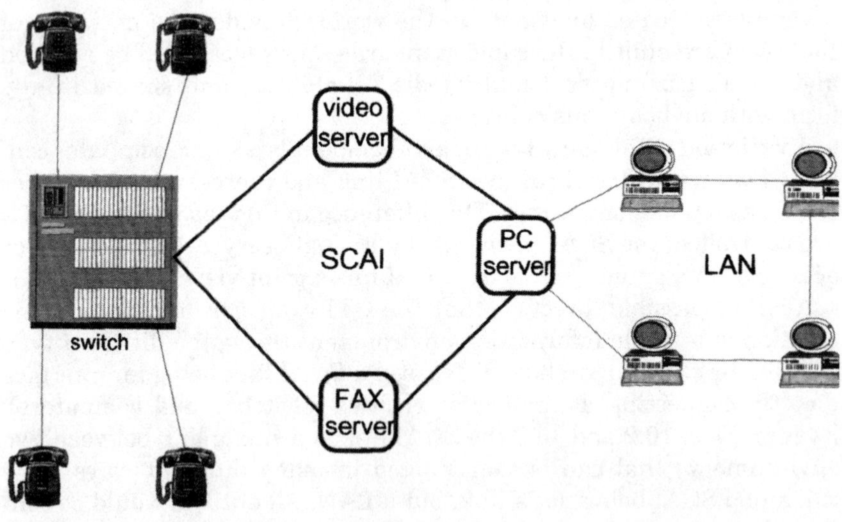

Figure 10.4 SCAI as a LAN.

partners in the SCAI environment. A user does not know about the specific arrangement, but benefits from it in the form of the ability to access telephony, fax, video, and other possible services by using one device, for example, a PC.

A modification of this idea is the SCAI LAN connecting multiple PCs without the PC server, as in Fig. 10.5.

If any of the scenarios shown in Figs. 10.2, 10.4, or 10.5 becomes a reality, the SCAI standard may have to change to adapt to the new interconnection scheme. First, there would probably be a need for more messages (FEs) to perform additional functions. Second, some kind of a priority structure would have to be established to allow more critical applications a priority access to the resources. For example, if a voice call and a fax call are originated to the same destination by two different entities, someone has to decide who gets connected. What else would have to be changed or added depends on the specifics of the servers to be used and is very difficult to envision right now. However, the basic SCAI architecture should support these new developments. The SCAI independence from the transport mechanisms should allow the use of LAN protocols for SCAI information transport.

Regardless of the direction that CTI and SCAI evolution takes, one issue is definitely important: development of the standard API. We talked about the inability to model the data processing environment in Chap. 3. At the present time, the SCAI standard assumes an existence of an API as the representation of the data processing environment. The CTI evolution, however, requires the development of such standardized API. It would provide opportunities for applications writers to economically design software for applications that can be used on different computers. The standard API would also simplify access from an application to the SCAI environment, allowing applications design-

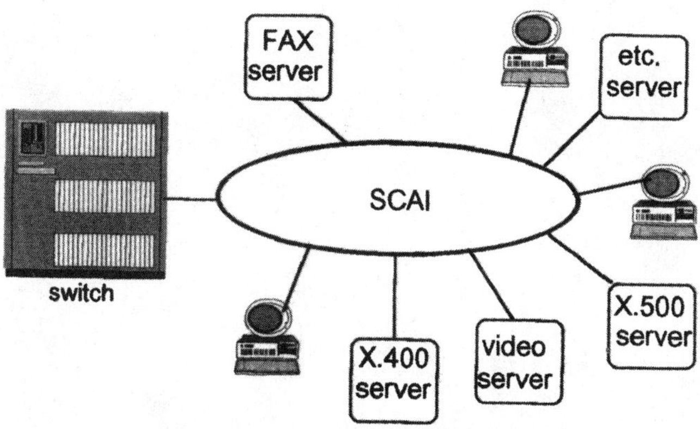

Figure 10.5 Another SCAI LAN alternative.

ers to concentrate on the details of the applications. Uniformity and portability of SCAI software would allow installation of the SCAI applications in a large number of users' computers, which in turn would affect the price and allow even more users to acquire applications with SCAI access.

The importance of the API for CTI is currently recognized in the industry. Unfortunately, no standard for API is yet being developed or planned. If the CTI world is to move forward, API standards will be needed.

Chapter

11

Implementation Examples

So far in this book we have discussed the SCAI standard, its usefulness and functionality, its open issues, and its future. The standard is a solution to CTI that provides great benefits for the industry and users. We talked about the benefits before and will not repeat the discussion here. In this chapter we will cover the CTI implementations that are not based on the SCAI standard, but on the vision that a specific manufacturer of the CTI equipment had at the time its equipment was developed. These implementations are close to SCAI in functionality and the type of applications that they intend to support. They are different in the specifics of their implementations: the messages and parameters, some functions, the overall capabilities, and the setup.*

These examples of real implementations demonstrate the pre-SCAI technology and show how manufacturers responded to the market needs without the benefit of a standard. Two PBX and one computer implementation from three different manufacturers are discussed. The implementations existed for a number of years and were developed separately, but were intended for the same market, yet, expectedly, they are different. The result is that if a computer is to be connected to different switches, it has to be equipped with different software packages to handle each specific switch.

There are many different switch and computer manufacturers of CTI equipment. The purpose of this chapter is to overview the technology based on several manufacturers' implementations. The manufacturers whose equipment is discussed in this chapter are IBM, NEC, and Inte-

* There is also a possible difference in terminology with the result that the same term may mean different things in SCAI and in a specific manufacturer's example of implementation.

com. IBM is, of course, a computer maker, and NEC and Intecom are PBX makers.* There is no information at the time of this writing regarding these manufacturers' plans to migrate their implementations toward the SCAI standard, and this is not the point of the presentation anyway.

I would like to point out that the features and capabilities† of the discussed products are presented based on the information supplied by the manufacturers of these products. I do not attempt to make any value judgment regarding them. Specific questions regarding any product, its availability, detail characteristics, and evolution, should be directed to the manufacturer of that product.

11.1 NEC PBX

The NEC CTI product is called Open Application Interface (OAI). It has been available since 1988. The OAI is based on NEC's NEAX 2400 PBX. Architecturally, OAI is the same as SCAI—it is an interface between a switch and an application computer. The OAI's purpose is to coordinate processes in telecommunications and computing environments, which is also the same as that of SCAI. NEC sees the OAI as a general interface that can be used for different reasons depending on the customers' needs.

The question may be asked: What does it mean to install OAI? As a matter of fact, it is the same question regardless of which CTI product is considered. Installing any CTI product usually translates into acquiring a switch which is equipped with hardware and software to handle CTI processing. Another item needed for OAI to become operational is, of course, a computer. We will discuss the computer side of this story when IBM's product is presented later in this chapter. If a switch—let's say a PBX—has already been installed for other reasons, it can be expanded to support the CTI functionality by acquiring additional hardware and software to run, in NEC's case, OAI.

The OAI applications are grouped into three categories:

- *Real-time Database Extensions.* These applications allow access to expanded databases, thereby providing new sophisticated services.
- *Interactive Telephone Terminal.* These applications provide interactive telephony-oriented services.

---

* NEC does make all kinds of switching equipment for public and private networks, but here we only look at NEC's PBX product.

† As an exercise, you can compare how these relate to those standardized as part of SCAI.

- *Application Control of Call Processing.* These applications provide services that require coordination with call processing.

TCP/IP and X.25 are two transport mechanisms currently specified for the link connecting the NEAX 2400 PBX with a computer. Selection of either depends on local circumstances.

11.1.1 Applications

Real-time database extensions. The following applications belong in this category:

- *Roamer.* This is a specialized call forwarding feature. When you are in the environment served by the NEAX 2400 with OAI, you can request forwarding of calls to a specific telephone by entering appropriate information at this telephone. This feature allows easy forwarding of calls to different locations as the person moves from one location to another. Without this feature, call forwarding is activated at the person's telephone, which means that every time a "forwarding to" destination changes, the person must go back to his/her office to reprogram call forwarding. With a roamer feature, the same person can activate call forwarding to each telephone on his/her route by entering a personal identification number at each telephone. The PBX (switch) recognizes the ID of the telephone at which the request was made, and the computer translates a personal ID into a telephone number for which calls are to be forwarded. The coordination of information between a PBX and a computer produces the desired result.
- *Code Master.* This application allows the expanded use of authorization codes to activate different services, such as making calls.
- *Name Display.* This application displays the name of a caller or a person the call is placed to. When this feature is activated, the computer supplies the necessary information for the display.

Interactive telephone terminal. Four applications are in this category:

- *Time Clock.* This application provides the ability for employees to register (check in/check out) at designated phones. The employees' input is received by a PBX and sent to a computer over OAI for storage and further use.
- *Asset Master.* This application allows access to reservation information (for conference rooms, etc.) over a multifunction phone. Again, OAI coordinates processes in a PBX and a computer to achieve this.

- *Directory Dialer.* This is an expanded version of abbreviated dialing. It allows storage of destination telephone numbers and their automatic use and display.
- *Second Party Alert.* This application allows the identification of the caller on another line without interrupting the current call. The calling number is displayed when an appropriate button is pressed. If Name Display is also activated, the name of the caller is displayed.

Application-directed Call Processing. The following applications are in this category:

- *Operator Assistance.* This application allows an operator (secretary, etc.) to process calls for internal users. An operator can access the personal directory of each user; call-related information is displayed at the terminal.
- *Guardian.* This application guards the use of telephones for long distance calling. It monitors long distance call attempts, collects data, detects invalid call attempts, and, when necessary, denies service.
- *Emergency 911.* This application displays information regarding the telephone extensions generating emergency calls. It also makes available information regarding emergency-response agencies based on which actions can be taken.
- *Security Monitor.* This application allows an operator to monitor the status of a telephone line and, at the user's request, listen in to the conversation.
- *Attendant Console Systems.* This family of applications provides enhanced attendant capabilities, including on-screen displays, administrative features, and messaging capabilities. The following specific application packages are currently available:
  —Medical Center System—designed specifically for hospital environment
  —Attendant Controller System—designed for hotel/motel environment
  —Attendant Management System—designed for internal organizational needs of any department or organization

11.1.2 Functionality

The NEAX 2400 PBX has to be equipped with OAI software to support the applications discussed in the previous chapter. This software contains specific OAI functions which are invoked when a particular need exists. The functions are grouped into three categories:

1. Call Functions—functions related to call processing
   - Make Call (voice or data)
   - Release Call (voice or data)
   - Transfer Call
   - Create Conference
   - Monitor Connection
   - Join two connected points
   - Connect a call at a monitored number
   - Force connect to an "off-hook" extension
   - Retrieve a ringing call
2. Event Functions—functions related to reporting of occurred events
   - Trunk connected (incoming/outgoing)
   - Trunk released (incoming/outgoing)
   - Monitored number connected
   - Monitored number transferred
   - Monitored number forwarded
   - Station on hook
   - Station off hook
   - Station answer
   - Station off-hook alarm
3. Control/Data Functions—functions related to the administration of the system
   - Switch-related:
     Call Authorization
     Move Logical Extension
     Alter Extension Restriction
     Alter Extension Features
     Alter Call Routing Patterns
     Report External Call Data
   - Terminal-related:
     Enter/Exit Terminal Mode
     Write to a device (display, lights, chimes)
     Read Keypad Input

An important functional element of OAI is the Application Manager. It is a menu-driven software package that is used, as its name suggests, to manage the OAI configurations. The Application Manager is described as supporting the following tasks:

- Creation and editing of configuration files that include key codes, program parameters, required facilities, and so on.
- Creation and editing of system generation files that include all software parameters (e.g., number of buffers and queues, number of applications, etc.).
- Initialization of application processes.
- Activation of system monitor.
- Creation and editing of process control files that designate automated initialization and termination of applications.
- Monitoring and control of system status and commands.
- Execution of non-OAI user programs.
- Management of functionality which includes debugging, application testing, file archiving, process simulation, and so on.
- Other management functions.

11.1.3 OAI customization

Whether a new NEAX 2400 PBX with OAI is purchased or an OAI system is added to the existing one, two basic options for OAI software exist. The first option is to acquire all or some applications that have been developed for OAI. In this case, only installation of the system and its provisioning is required to begin operations. This option may be suitable for those who require specific applications already developed in a particular manner.

The second option is to acquire the basic software package that allows development of different applications. In this case, applications would be developed by a user of equipment or by the third party. In either case, the applications would be developed to satisfy specific needs of customers.

11.2 Intecom PBX

The Intecom CTI product is also called OAI. It is based on the Intecom's family of PBXs—IBX and Telari. IBX is the larger of the two PBXs, and both can accommodate OAI functionality. When IBX is referenced in this chapter, both IBX and Telari are meant. Architecturally, the Intecom's OAI is the same as SCAI and is a link between an IBX and a general purpose computer. Intecom stresses the OAI as a tool for solving business problems.

11.2.1 Applications

The IBX OAI contains a set of functions that are used by applications depending on their needs. The functions will be described in the next chapter, while here the potential applications are presented. Intecom has produced application notes to describe the applications that can be supported by the OAI. These notes can be acquired directly from Intecom. Applications other than those described in the notes can be developed, provided they use functions available in the OAI system.

- *Data Security.* This application provides user validation and tracking for access to host computers. When a data call is to be made, the request is passed to a computer which validates the request and itself places a call to a user.
- *PC Based Stored Number Dialer.* This application provides the ability to initiate calls from the computer based on supplied numbers or on an available list of destination numbers.*
- *Message Waiting Indication.* This application provides the ability for any user to activate/deactivate his or her message waiting indicator. Certain users (e.g., a receptionist) would be allowed to perform these actions for some other people in the organization.
- *Travel Reservation System.* This application provides travel agents with the ability to better serve their customers through the use of OAI. When a customer calls the agency, an IBX supplies a computer with calling information based on which customer's profile is retrieved and displayed at the agent's terminal.
- *Hotel Room Status.* This application allows maids and maintenance personnel to update the status of hotel/motel rooms through the telephones located in the affected rooms.
- *Message Center.* This application provides a mechanism for a group of operators to accept and distribute messages efficiently and in a personal manner.
- *Parking Lot Security.* This application pinpoints, via a graphic display, the exact location from which an emergency call is originated.
- *Campus Emergency.* This application provides an emergency operator with a caller's location, additional information about the caller, and information necessary for proceeding with the emergency call (e.g., who to call in case of emergency).

---

* Although Intecom's literature specifies that a PC is used to store the destination numbers, I suspect that any kind of a computer can be used for this application.

- *Time Billing.* This application provides an automatic mechanism for tracking employees' involvement with different projects.
- *Electronic Mail Notification.* This application provides users of the electronic mail system with information about "unread" mail and other aspects of electronic mail. This application can handle two situations: (1) when OAI and electronic mail systems reside on the same computer and (2) when multiple computers are involved.

11.2.2 Functionality

Intecom's OAI functions are grouped into two categories. The first category is called Call Processing Services and contains functions related to call processing, such as making calls, reporting progress events, transferring calls, and activating various features at the telephone. The second category is called System Administration Services and contains functions that provide access to the IBX's statistical information. This information can be used to produce customized reports or as an input to other applications.

Call Processing Services. Several groups of functions belong in this category. It would be easier if we introduce them separately.

- *Application Requested Calls.* A computer can request the following tasks to be performed by the IBX:

  Make a call (voice or data)

  Transfer a party

  Disconnect a call

  Activate/Deactivate a station* feature on behalf of a user. The current set of features that can be activated/deactivated is as follows: Hold, Conference, Call Park, Do Not Disturb, Callback, Answer/Release, Silent Monitor, Intercom, Page, Last Number Redial, Call Forward All, Privacy Release, Busy Override, Group Listen, Hot-line, Call Pickup, Speaker On/Off. In addition, at the attendant console, Join and Disconnect features can be affected.
- *Activity Indication.* An application can instruct the IBX to notify a user about certain events. This is accomplished via a flickering feature lamp; cancellation of the notification is performed when certain criteria have been satisfied (e.g., time-out, user retrieved information, etc.).

---

* A station is a telephone.

- *Display Text.* An application can request the display of certain information at the user's display phone.
- *Feature Initiation.* An IBX user can communicate with an application residing in a computer (connected over OAI) via a sequence of dialed digits. The dialed digits are collected by the IBX and passed to the computer over OAI. The application takes actions based on the information supplied by the dialed digits.
- *Call Information (Group-based).* An application receives reports regarding the progress of calls related to groups: trunk groups, ACD groups, attendant groups. The reports include information regarding circumstances surrounding the call, involved parties, and so on.
- *Call Information (Individual Facilities).* An application receives reports of events visible at the individual facilities: trunks, lines, stations, and so on. Examples of events that are reported: ringing, off-hook, call on hold, etc.

System Administration Services. These services allow applications to access certain tables in the IBX. The purposes for the access are based on the applications' needs and can be collecting data, modifying data, and creating/deleting data. The parameters that can be accessed are Directory Numbers, Ports, Authorization Numbers, Class of Service, Terminal Type, Agent ID, and so on. An application can schedule the task of collecting or updating database tables at the time which is most convenient—either when the call processing activity is insignificant or when it is necessary.

11.2.3 OAI customization

Intecom's OAI is also designed with the development of new applications in mind. According to Intecom, applications that are tailored to specific environments and customers can be developed as extensions of the basic OAI platform.

Access to the OAI development laboratory can be arranged either on a dial-up or on-site basis. The laboratory enables testing of customers' applications in a real live environment. An application development toolkit is also supplied. The toolkit is a set of programs that help in the development of applications. Included in the toolkit is an OAI API. Intecom supplies a documentation package which includes reference manuals, developer's guides, and the OAI tester/emulator with the associated documentation. In addition, a training course in OAI is available.

11.3 IBM

IBM's CTI product is called CallPath. CallPath is a service architecture supported by a variety of IBM systems—System/390 and System/370, PS/2, RISC System/6000, Application System/400. It means that any of the listed systems can be used as CTI hosts (computers). The IBM documentation identifies a number of switching systems that can currently work with CallPath, such as those from AT&T, NEC, Northern Telecom, ROLM, and Siemens, but in reality, any PBX or a public network switch can be accommodated if the need arises.

11.3.1 CallPath architecture

The discussion of the CallPath architecture is necessary to understand how IBM has engineered its CTI system and how it can be made functionally universal across IBM products. Fig. 11.1 shows how various functional CallPath components work together.

Three types of units are shown in the figure: a computer, a switch server, and switches. The computer is an IBM machine which contains applications. The type and number of applications depends on the specific needs of the customer deploying this equipment. Also in the computer is a program which provides CTI communications functionality. This is shown as a CallPath Subsystem, which is specific for each type of computer, e.g., CallPath/6000 for RISC System 6000. The subsystem's functions allow the applications program to communicate with a switch. Between an application program and a CallPath Subsystem is an API,* which allows different types of application programs to communicate with a Subsystem.†

The subsystem communicates with a switch, but, because different switches currently use different nonstandard protocols, IBM has designed a protocol conversion unit—SwitchServer/2. This unit resides in the IBM PS/2 system and allows a host computer to be free of concerns of specific switch protocols. It maps any switch protocol to the IBM protocol and, therefore, allows an application developer not to be concerned with the specifics of a switch protocol.

CallPath works in the following way. An application running in a host computer needs to communicate with a switch. It issues a program call to the API which is passed to the CallPath Subsystem. The subsystem issues a required message which is transmitted to the SwitchServer, mapped to a switch-specific message, and transmitted

* See Chap. 3 for more on APIs.

† If we would try to map this functional architecture to that of the SCAI standard, we could say that a subsystem performs the functions of the SCAI ASE and associated ASEs (ROSE, ACSE, and Security, depending on the situation).

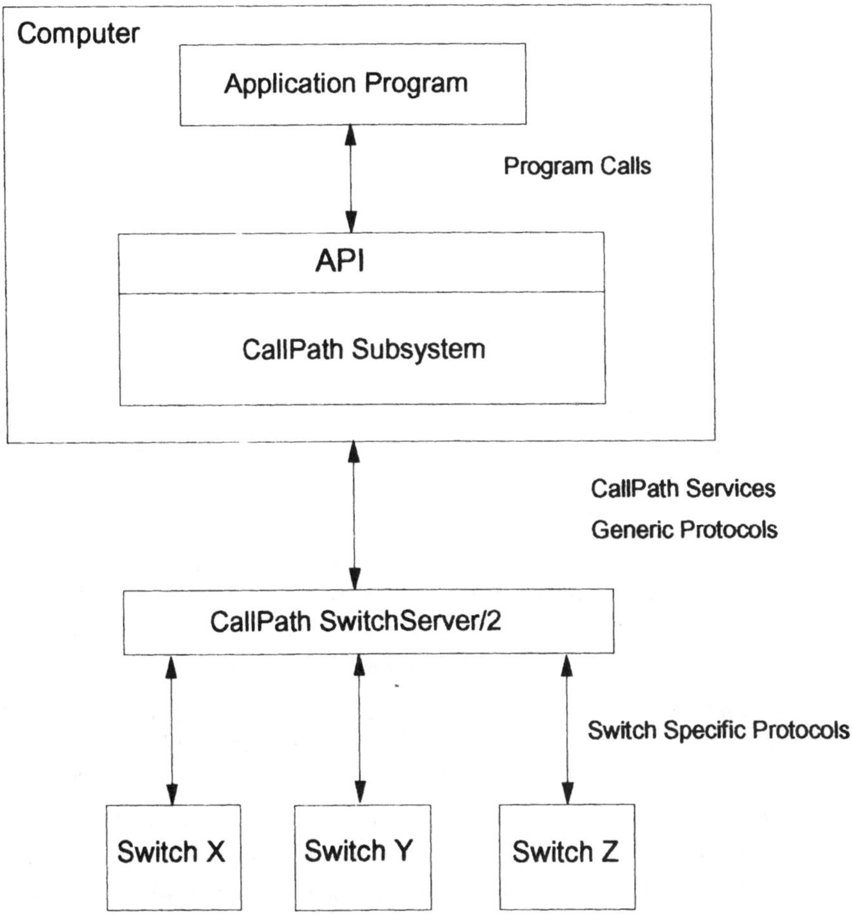

Figure 11.1 CallPath functional architecture.

across the link to a switch. A message coming from the switch to the host would experience the opposite transformations. Let's look at the physical CallPath architecture shown in Fig. 11.2.

The transport mechanism used between a computer and a switch server is usually one of the following: X.25, SDLC, IBM Token Ring, or Ethernet. The transport mechanism between a switch server and a switch is anything supported by both PS/2 and a switch (X.25, IBM protocols, etc.).

11.3.2 CallPath functionality

The services that CallPath provides are targeted to support a wide range of applications. CallPath can be viewed as a collection of func-

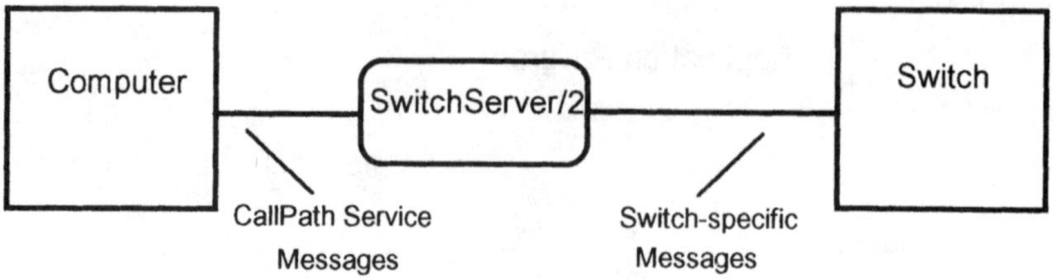

Figure 11.2 CallPath physical architecture.

tions that are used as required in a specific customer environment. However, IBM, like other manufacturers of CTI equipment, identifies a set of services that can be directly supported by CallPath functions. These services are as follows:

- *Intelligent Answering.* Retrieval of customer's (caller's) data from the computer file based on the calling or called number. The voice call and associated data are delivered simultaneously to an agent.
- *Coordinated Voice/Data Transfer.* Transfer of a voice call and associated data (screen information) to another person.
- *Consultation.* The ability to involve a third party (e.g., a supervisor) in the discussion between a customer and an agent. The third party is capable of accessing the same data which is used by the agent.
- *Intelligent Dialing.* Automatic dialing of telephone numbers for services like telemarketing or payment collections.
- *Personal Services.* A class of services performing conferencing and hold/retrieve functions.
- *Transfer Load Balancing.* Transfer of calls to the least-busy agent group regardless of its geographic location.
- *Voice and Data Activity Log.* Collection of pertinent call-related data that can help improve operations (e.g., statistics).

A number of functions are defined to support these and other services. To invoke any of the CallPath functions a program call is issued to the API and is mapped onto a specific routine in the appropriate subsystem. Following are the currently defined CallPath functions (in alphabetical order):

- Add Party—adding a party to a call.
- Alternate Call—alternating between two calls which are in turn active and on hold.

- Answer Call—answering a call on behalf of the called party.
- Conference Call—merging two calls to create a conference.
- Create Program Name—creating a unique program name in a subsystem.
- Delete Call Profile—erasing a previously created call profile.
- Disconnect—removing a party from a call.
- Extend Call—placing an existing call on hold and attempting a connection to another party.
- Hold Call—placing a call on hold on behalf of another party.
- Identify Program Name—identifying the program name to the CallPath services subsystem.
- Immediate Transfer—transferring an existing call to another party without placing the call on hold.
- Initialize Call Profile—initializing a program-specific call profile.
- Invoke Feature—invoking a feature (specified by a parameter) for a telephone associated with a specified party.
- Make Call—Initiating a call.
- Monitor—registering interest to receive specified call progress event messages for a specified resource (e.g., a directory number*).
- Monitor System Messages—registering interest in receiving system messages.
- Park Call—parking a call.
- Pick Call—picking a call.
- Provide Indicator—providing an indicator to a telephone associated with a specified party.
- Query ACD Information—soliciting ACD statistics.
- Query Call Parties—soliciting information about the parties involved in telephone calls.
- Query Hardware Status—soliciting information about various types of hardware (e.g., trunks).
- Query Party Status—soliciting information about a specific party.
- Receive—receiving messages.†

---

* A number by which a specific telephone or other telecommunication device is identified.

† This function, when invoked, allows the messages generated by a switch to be received by the application. The internally generated messages would also be received.

- Redirect Call—changing the designation of an incoming call.
- Register Ownership—establishing ownership of a specified resource, e.g., a directory number.
- Reject Call—rejecting a call.
- Release Program Name—requesting to discontinue usage of a specific program name.
- Retrieve Call—retrieving a call placed on hold.
- Return Control—returning control of a call to a switch.
- Send Device Data—transmitting telephone digits on behalf of a specified party.
- Set Add Party Directions—setting the *add-party-direction* call profile characteristic.
- Set Automatic Answer—setting the *automatic-answer* call profile characteristic.
- Set Billed Party—setting the *billed-party* call profile characteristic.
- Set Call Type—setting the *call-type* call profile characteristic.
- Set Called Party Alerting Time—setting the *called-party-alerting-time* call profile characteristic.
- Set Disconnect Execution Option—setting the *disconnect-execution-option* call profile characteristic.
- Set Extend Purpose—setting the *extend-purpose* call profile characteristic.
- Set Held Call Connection—setting the *held-call-connection* call profile characteristic.
- Set Held Call Recording—setting the *held-call-recording* call profile characteristic.
- Set Holding Party Callback—setting the *holding-party callback* call profile characteristic.
- Set Holding Party Calling—setting the *holding-party-calling* call profile characteristic.
- Set Make Call DND Override—setting the *make-call-DND-override* call profile characteristic.
- Set Make Call Forward Override—setting the *make-call-forward-override* call profile characteristic.
- Set Make Call Notification—setting the *make-call-notification* call profile characteristic.

- Set Make Call Party Order—setting the *make-call-party-order* call profile characteristic.
- Set Monitor Tracking—setting the *monitor-tracking* call profile characteristic.
- Set Parking Party Calling—setting the *parking-party-calling* call profile characteristic.
- Set Party Specification—setting the *party-specification* call profile characteristic.
- Set Remaining Party Option—setting the *remaining-party-option* call profile characteristic.
- Set Return Response—setting the *return-response* call profile characteristic.
- Transfer Call—requesting a transfer of a call to another party.
- Trigger—triggering an action in a switch when a specified event is encountered.

Some of the functions generate messages to a switch (e.g., Make Call, Redirect Call), while other functions are performed internally in a computer (e.g., Identify Program Name).

11.3.3 CallPath customization

Two aspects of customization can be addressed. The first aspect is the use of CallPath functions by different applications. When a specific application is to be developed using IBM computers, all functions discussed in the previous chapter are available to the application designers. How the functions are used depends on the developer, or, as we should say, depends on the requirements of a particular application.

The second aspect of customization is switch-dependent. IBM already developed switch-specific packages for a number of switches available on the market. When such a package is developed, it is tested with a specific switch. A communications package for any switch can be developed, allowing any PBX or public network switch to be connected to IBM computers.

IBM supplies to the users the CallPath services documentation. This documentation includes a description of programming, installation, and problem determination guides for each specific computer (e.g., PS/2, System/370) and guides for using CallPath with specific switching systems.

Chapter

12

T1 Standards Process

Committee T1 (Telecommunications One) was created at the time of the AT&T divestiture with the intent to define domestic United States standards in telecommunications. Since its inception in the beginning of 1984, T1 has been involved with all technologies that are considered important by the telecommunications industry. It produced and is still producing standards for ISDN, Common Channel Signaling, and Intelligent Networks, to name just a few. T1 is accredited by ANSI, which means that T1 draft standards are approved to be American national standards through the ANSI process. This process assures fairness in deliberations, broad and equal participation by various industry segments, and publication of standards in a specific common format.

Participation in various organizations of the Committee T1 requires resources, and large companies are in a better position to have such resources. As a result, large companies like AT&T, Bellcore, IBM, Northern Telecom, etc., introduce the largest number of contributions to progress draft standards in T1. However, participation by smaller companies not only benefits the industry by providing additional input to the process, but can provide these companies with benefits. The benefits would be the same as for the large companies: advancing a standard in a specific direction, getting familiar with different views in the industry on a given subject, staying abreast with newest standards, and so on. Even the impact that a small company can have on standards development is possibly significant. A standard is being shaped by individuals; they represent various companies, of course, but they can only impact the process proportionally to their abilities. An individual possessing expertise in a subject and great negotiating skills usually produces the most impact, and any company can be represented by someone like that. This is not unique to Committee T1, but this is where I saw great examples of the impact that an individual from any company can have on the development of an important standard.

While it would be beneficial for the industry if a greater number of companies were actively participating in T1, the point of the preceding paragraph is to stress the other aspect of the participation issue in T1: the types of participants. Most of the active standards developers are manufacturers and network providers. It would be extremely helpful if so-called users participated also. Users are not necessarily small companies. On the contrary, banks, insurance companies, and government agencies are rather large enterprises. Unfortunately, their philosophy seems to be that the technical standards should be developed by experts in specific technologies. The problem is that usually there are many options for technical solutions and many times only a user can state which is the most useful option. This is a great and underutilized tool that the users have—the ability to influence the directions and contents of standards—and hopefully it will be used more often in the future. Understanding the rules and procedures of T1 as well as the ways to impact the process would greatly benefit those wishing to participate in the work of any group within Committee T1.

12.1 T1 Structure

Committee T1 consists of Technical Subcommittees (TSCs), the number and type of which varies based on current needs in the industry. At the time of this writing (the beginning of 1994), six TSCs were operational (see Fig. 12.1).

Each TSC has its own charter and is responsible for standards in a specific area. When a TSC fulfills its duties, it is disbanded. Over the years there were also some changes, inspired by specific conditions, resulting in reshuffling projects among TSCs. For example, current T1S1 is responsible for projects that used to be covered by T1D1 and T1X1. Other TSCs that used to exist are T1C1 and T1Q1. The responsibilities of the current TSCs are as follows:

- *T1A1—Performance and Signal Processing.* This TSC is responsible for standards in performance and the processing of voice, data, image, and video signals. Specific areas of involvement are voice net-

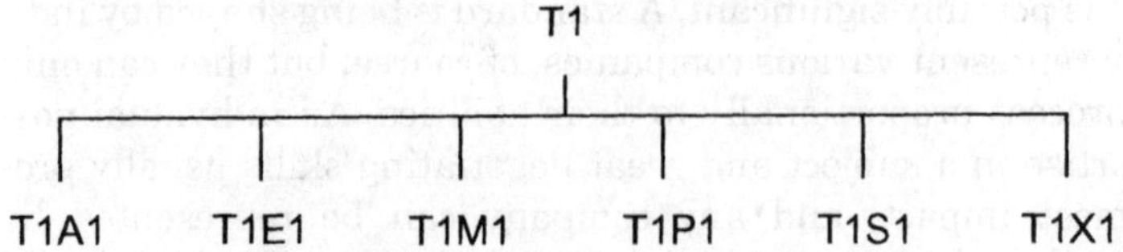

Figure 12.1 Structure of Committee T1.

work performance, packet data performance, ISDN performance, specialized signal processing, network survivability performance, and audiovisual communications coding and performance.

- *T1E1—Interfaces, Power, and Protection of Networks.* This TSC develops standards for network interfaces, power, and electrical and physical protection. The interface standards developed by this TSC are limited to the physical aspects of the interfaces, i.e., their electromagnetic, optical, and mechanical characteristics. Specific areas of involvement are different types of interfaces (analog, wireless, wideband), connectors, power systems, electrical protection, physical protection.
- *T1M1—Management and Maintenance.* This TSC develops standards for network operations, administration, maintenance, and provisioning.
- *T1P1—Program Management and System Engineering.* This TSC was initially envisioned as being responsible for standards planning and program management. The first task the TSC took on was PCS.* However, the work on PCS transcended the program management aspects and currently includes technical work on PCS. Specifically, the TSC is involved with PCS requirements, modeling, definitions, addressing, service descriptions, and terminology.
- *T1S1—Services, Architecture, and Signaling.* This TSC develops standards for ISDN, Common Channel Signaling (SS7), Broadband ISDN, and Intelligent Networks. It is concerned with services, architectures, and signaling protocols for the foregoing technological platforms. The SCAI standard has been developed by this TSC.
- *T1X1—Digital Hierarchy and Synchronization.* This TSC develops standards for network transport signals. Specifically, it is involved with issues of synchronization, digital hierarchy, and metallic and optical interfaces.

Each TSC meets independently and has its own chair and vice-chair. The entire Committee T1 is governed by the bylaws and decisions made at the committee's meetings. Between the meetings, the Advisory Group (T1AG) supervises the affairs of the committee. T1AG is composed of the committee officers: T1 chair and vice-chair, Representatives of each membership categories, and all TSCs' officers, who can participate in the deliberations, but cannot vote. In simple terms, T1AG governs the affairs of T1 and seeks approval from Committee T1 when necessary.

* Personal Communications Services.

The membership categories were created to assure proper representation in the committee by different players of the industry. There are four categories:

- Exchange interests (local telephone companies)
- Interexchange interests (long distance telephone companies)
- Manufacturing interests
- User and general interests

The division into categories has more significance than just classification on paper. Certain decisions cannot be approved if a single membership category objects to it. For example, a standard that received sufficient general vote may not be approved if all members belonging to one category (e.g., users) vote "No." This guarantees fair attention to everybody's needs.

Another unit of Committee T1 is T1 Secretariat. It performs functions associated with the administration of the committee, external interface, and publishing of the approved documents. A representative from the Secretariat is always present at the T1AG meetings. If you are interested in obtaining any formal descriptions of T1 procedures and bylaws, contact the T1 Secretariat:

T1 Secretariat
c/o ATIS*
1200 G Street, N.W., Suite 500
Washington, D.C. 20005
Telephone: 202-434-8845
Fax: 202-393-5453

12.2 TSC Structure

Each TSC is composed of Working Groups, which, in turn, are composed of Subworking Groups and ad hoc groups. Although each TSC creates its own structure, the rules of their creation and behavior are, of course, the same. Another important point is that TSCs work based on the approved projects (e.g., SS7 TCAP protocol). The Working Groups are usually given the responsibilities for individual projects (although it does not have to be that way), and each Working Group decides how to better accomplish the project; that is, it creates specific subworking groups or ad hoc groups to do that.

A TSC elects a Chair and Vice-Chair of each Working Group (WG). Chairs of Subworking Groups are appointed by a Working Group Chair

* Alliance for Telecommunications Industry Solutions, formerly known as ECSA.

with the approval of the Working Group. Convenors of ad hoc groups are also appointed by the WG Chair.

A Subworking Group (SWG) is created for the duration of the project(s) assigned to it. Usually, SWGs exist a long time because standards projects are not accomplished in months, but in years. An ad hoc group has to be approved at each WG meeting. An ad hoc group considers an issue assigned to it, reports to the WG, and disbands. At the next WG meeting, the same ad hoc group may be re-created, usually with the same convenor. Ad hoc groups are typically small groups considering issues that seem to be of less importance to the industry at the moment. SWGs are actually the level at which all the work of defining and negotiating a standard is performed. An SWG is responsible for a particular standard and reports its accomplishments to the WG at each meeting.

12.2.1 T1S1 structure

To give an example of a TSC structure and at the same time to present the environment in which the SCAI standard has been developed, let's discuss one of the TSCs: T1S1.

T1S1 is involved with all the new technological platforms the industry is developing: ISDN, IN, etc. It consists of three Working Groups*:

- T1S1.1†—ISDN Architecture, Services, and Signaling Protocols
- T1S1.3—Common Channel Signaling
- T1S1.5—Broadband ISDN

Let's now look at the composition of T1S1.1. It currently (beginning of 1994) consists of five subworking groups: SBSD (Supplementary and Bearer Services Descriptions) SWG, DSS1 Services SWG, Circuit Mode SWG, IN SWG, and ISDN Networking subworking group. There also has been an ad hoc group looking at ISDN terminal adaption issues, but it does not meet any longer.

The SBSD SWG defines ISDN services—so-called Stage 1. These are service descriptions from the point of view of a user and are necessary for defining a protocol (which, using the same ISDN terminology, is called Stage 3 description). Examples of ISDN services are Basic Call, Call Conferencing, Call Forwarding, and so forth.

---

* Two other Working Groups used to be a part of T1S1.1, but do not exist any longer—T1S1.2 and T1S1.4. T1S1.2 was responsible for the development of ISDN signaling protocols and was just recently disbanded, its responsibilities split between T1S1.1 and T1S1.5. T1S1.4 was responsible for numbering.

† Pronounced "tee one es one dot one."

The DSS1 Services SWG defines the protocols to support ISDN supplementary services, such as Call Waiting, Call Forwarding, and so forth. The Circuit Mode SWG defines signaling protocols for circuit-switched capabilities. These two subworking groups define the protocols based on the service descriptions created by the SBSD SWG.

The IN SWG defines Intelligent Network standards. The scope of work for the IN SWG has been, until mid-1993, to contribute to the IN standards process in ITU-T. In fact, the ITU-T IN standards published in 1993 are largely influenced by the work in this SWG. In the middle of 1993, T1S1 approved an IN standards project to define the American national standard.

The ISDN Networking Subworking Group is actually a misnomer in terms of the name. This is the group that developed the SCAI standard. The reason it has a strange name is historical (like so many things in standards). The SWG was organized to address the issues of private networking in ISDN. The issues of private network architectures, services, and protocols were discussed and the Technical Report was issued in March of 1992. The SCAI project started in parallel with a private networking project and became the single task of the group when interest in the private networking issues expired.*

12.3 T1 Procedures

Committee T1 produces two types of documents: standards and technical reports. Standards have to be approved by ANSI and become American national standards. Technical reports are approved by T1 and constitute informational reports on the state of agreements in a specific area. Technical reports are produced when it is believed that the information would be of benefit to the industry and the public, but the nature of information does not lend itself to a formal standard. For example, a user's guide associated with a particular standard can be produced this way. Another reason to produce a technical report is to inform the industry and the public of the current work status in a specific area.

All output of the committee is produced by its membership (no surprise here!). Membership costs money and, therefore, can be too expensive for small companies. However, there are two levels of membership in various organizations of T1: one that costs money and another that is free. Participation in the meetings and decisions of T1 and the TSCs is

* For completeness' sake, I need to mention that the IN project in the beginning was also assigned to the ISDN Networking SWG. So the SWG had three parallel projects: SCAI, IN, and private networking. When the IN project grew significantly, a new SWG was created to focus just on it.

by paying members. However, anybody can participate in the meetings of Working Groups and their subgroups! At this working level, where actual work on standards is performed, any company can participate without paying any dues. These participants receive the same documentation as any formal members of TSCs; they can contribute to the process, and their opinion counts in the definition of consensus. And this is a very important point in understanding the process in T1. While T1 and TSCs make decisions based on votes by members, WGs and SWGs operate on the basis of consensus. Consensus is called by the Chair of a group based on comments and discussion. There is no hard definition of consensus except that the minority should be relatively small and that the minority has an ample chance to express its view.

The consequence of such liberal membership rules at the working level is the ability by any company however small to participate in the process and significantly influence the process. There are plenty of instances when a small company affected the outcome of the standard through participation at the working level. Any company, large or small, can affect the process in two ways: by participating in the discussions and by bringing contributions to the meetings. Both ways are important, but it is impossible to influence a standard in a specific way without contributions.

Contributions are technical or organizational papers written by the participating companies to propose and advocate a specific point of view. A contribution can suggest a technical solution, propose a different organization of work, critique existing agreements, raise questions, or the like. A contribution is submitted to a specific Working Group. The WG then distributes contributions to appropriate SWGs or ad hoc groups, or considers them at its plenary meeting. By submitting I do not mean mailing the contributions to some office. Although this is allowed, it never happens because of the time constraints imposed on the distribution of the contributions through the mail. The contributions are submitted as follows:

1. A company prepares a contribution to a specific WG and receives a number, usually from the WG Chair or secretary. This number identifies the contribution for filing and recording purposes. For example, a contribution can be numbered T1S1.1/93-038. The meaning of the number is described in Fig. 12.2.

   The contribution number is printed on the title page of the contribution, which, in addition to the number, contains the following information:

   - Contribution title
   - Source of the contribution (company)
   - Contact name and telephone number

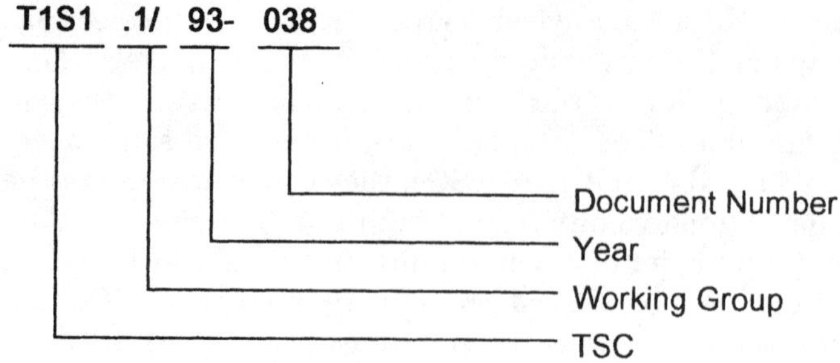

Figure 12.2 Contribution number.

- Project number under which the contribution is to be discussed
- Working Group name and, possibly, appropriate SWG
- Date and place of the meeting at which the contribution is to be discussed

2. The contribution is brought in for the WG meeting. The number of copies necessary for the distribution at the meeting is specified by each WG and depends on its membership. One critical point here is that everybody present at the WG meeting must receive a copy. Hence, members are told to bring the number of copies, which is usually in excess of regular attendance.
3. The contribution is designated for a specific SWG or ad hoc group, or is discussed by the entire WG. In most cases, the author of the contribution knows which particular organization will discuss it and directs it to that group in the wording on the title page of the contribution. Chairs of appropriate groups are usually aware of the approximate number of contributions expected at a given meeting through personal contact with members.
4. The contribution is discussed in the appropriate organization and a decision regarding its disposition is made in the meeting minutes.

12.3.1 Approval process

Contributions drive the development of standards and are written against a specific project approved by T1. Although there have been cases when the work started before the project had been formally approved, the practice is discouraged and occurs less frequently.

All standards and technical reports are built using temporary holding documents, in most cases called *Working Documents*. An editor is assigned for each working document and is responsible for recording text and figures as they were agreed on by the group. Editors are selected by SWGs and ad hoc groups. A Working Document contains agreements made by the group in the process of discussing contributions. It is the responsibility of a group's Chair to provide meeting agendas that include the sequence in which the contributions are to be discussed. This is necessary because in a case when there are more contributions than is possible to discuss in the time allotted for the group, the higher priority contributions will be discussed. It is also the responsibility of the Chair to direct and focus the discussion, call consensus when appropriate, or stop the discussion if no agreement can be reached.

The agreements are recorded in the Working Document only when consensus is reached and called by the Chair. These agreements are revisited later and can be modified. However, unless a technical error is found, it is very difficult to make changes in the existing agreements. Usually, more support is needed to change an agreement than to have one in the first place. That is why experienced participants are cautious regarding the agreements they make, and this is where negotiating skills are critically important.

After working on a specific project for a while, a SWG decides that a draft standard (or technical report) is ready for approval. At that time the draft document is brought to a WG for approval. If the WG agrees that the draft is ready, and it also works by consensus, the draft is passed to the TSC for further approval. A TSC can approve the draft standard either by voting at the meeting or by letter ballot. In the great majority of cases, a letter ballot is issued to which TSC members respond. Responses can be of several types: "yes," "yes" with comments, "no," or "abstain." Even if the ballot receives a sufficient number of votes for approval, the "no" votes and comments submitted with "yes" votes have to be discussed and all efforts have to be made to satisfy the commenting companies. This delays the process because in the presence of a "no" vote or if significant changes were made to the draft standard, the document has to be balloted again.

When the TSC approves the draft, it is the turn of Committee T1 to repeat the process again. And again, if there are comments, they have to be discussed in the originating working group. After T1 approves the draft standard, ANSI issues it for public comment, after which the standard is approved. The entire process from the initial approval by an SWG to the ANSI approval can take about a year. It can also take longer if one or more companies intend to use every opportunity to deny a standard, and these cases do exist.

In today's environment, lack of standards in certain areas can inhibit the progress and vice versa. Many people understand that the standards process is too slow to react to the needs of the industry and society. Attempts are currently being made to streamline the process without sacrificing the principle of fairness. One proposal which is supposed to be tried soon, is to allow voting on a standard by a TSC and T1 at the same time. This would save several months and will be tried with standards that seem to be noncontroversial. Whether these attempts will be successful remains to be seen, but it is encouraging to see that changes are possible.

12.4 External Liaisons

Committee T1 and the TSCs have another important function—corresponding with other standards bodies. There are a number of domestic and international standards bodies that develop standards in the same or similar areas as T1. Cooperation among different standards bodies helps the process and also helps to save resources (sometimes!).

The most important external body to T1 is ITU-T. Various Study Groups (SG) of ITU-T develop international standards in the same areas as T1, for example, SS7, ISDN, Management protocol, and so on. The United States as a country participates in various study groups and contributes to them through the appropriate TSCs. For example, T1S1 sends contributions* to SG 11 on signaling protocols and IN, to SG 1 on services, to SG 13 (former SG XVIII) on Broadband ISDN. The contributions constitute the United States' positions to be advocated at the ITU-T meetings. For all practical purposes, development of many ITU-T standards is supported by various TSCs. It is extremely important because it is to the advantage of everybody involved to minimize the difference between the United States' and international standards.

Committee T1 has relations with many other external organizations. Among them are ISO, to which contributions are sent through appropriate domestic committees (e.g., X3S3 approves contributions to ISO/JTC1 SC6), NIUF, IILC, IEEE, ETSI (European Telecommunications Standards Institute), ECMA (European Computer Manufacturers Association—a private European consortium), TTC (Telecommunications Technology Committee—the Japanese Standards Organization), and other regional standards bodies. All this external activity contributes to the development of better, more widely accepted standards.

12.5 How Can You Join?

As we discussed before, one can either become a member of T1, or a member of one or more TSCs, or just a participant in one or more

* It is not technically correct to state that a TSC sends contributions to ITU-T. A TSC sends draft contributions to the U.S. State Department for approval because governments are members of ITU. However, in reality, the vast majority of contributions generated by TSCs are approved.

Working Groups. The type of membership depends on the particular needs. If resources allow, it is best to be a member of T1. The second best is to be a member of a specific TSC where the standard of interest is being developed. These kinds of memberships allow full participation and voting and best cover the needs. To become a paying member you can contact the T1 secretariat.

If resources do not allow becoming a full member, participation in a specific Working Group and Subworking Group can be very effective. Such a participant cannot vote, but can affect the process at the working level. If you desire such participation, the following steps should be taken*:

1. Call the T1 secretariat to find out which TSC is working on the project of interest to you.
2. Call the TSC to find out which WG is responsible for the project.
3. Call an official of the given WG (Chair, Vice-Chair, appropriate SWG Chair) to find out the meeting schedule.
4. Come to the next meeting and participate in the opening WG plenary (it is best to inform the TSC secretary that you are coming for the meeting; it is needed for planning purposes).
5. At the opening plenary of a WG, collect all the documents of interest to you (contributions, meeting schedule, etc.).
6. SWGs meet after the opening WG plenary according to the schedule agreed on by the WG; participate in the SWG meetings.
7. Attend the closing plenary of the WG to hear the results achieved by all SWGs and ad hoc groups.
8. Starting with the following WG meeting, plan to bring contributions, if you already understand what is required to advance the work in a way favorable to your interests; or you can participate in the discussions until it is clear what you can do to help the standard along.

Most of all, to be effective, be active. Sometimes even observing is helpful by raising the understanding of the subject, but the only way to shape the standard is to bring your ideas, present them, and convince other participants that this is the best solution for everybody.

GOOD LUCK!

---

* The steps for beginning your participation in the meetings at the working level are exactly the same, whether you are paying membership dues or not.

Appendix

A

Abbreviations

AC—Application Context

ACD—Automatic Call Distributor

ACSE—Association Control Service Element

AE—Application Entity

ANI—Automatic Number Identification

ANSI—American National Standards Institute

AP—Application Process

APDU—Application Protocol Data Unit

API—Application Programming Interface

ASE—Application Service Element

ATIS—Alliance for Telecommunications Industry Solutions

ATM—Asynchronous Transfer Mode

BER—Basic Encoding Rules

BRI—Basic Rate Interface

CCS—Common Channel Signaling

CRC—Cyclic Redundancy Check

DSS1—Digital Subscriber Signaling System Number 1

DTMF—Dual Tone Multi Frequency

ECSA—Exchange Carriers Standards Association

ESP—Enhanced Service Provider

FCC—Federal Communications Commission

FE—Functional Element

IILC—Information Industry Liaison Committee

IN—Intelligent Network

ISDN—Integrated Services Digital Network

ISO—International Standards Organization

ISUP—ISDN Services User Part

ITU—International Telecommunication Union

LAN—Local Area Network

MACF—Multiple Association Control Function

MAN—Metropolitan Area Network

MIS—Management Information Services

NIUF—North American ISDN Users Forum

ONA—Open Network Architecture

OSI—Open System Interconnection

PBX—Private Branch Exchange

PC—Personal Computer

PCI—Protocol Control Information

PDU—Protocol Data Unit

POTS—Plain Old Telephone Service

PRI—Primary Rate Interface

PVC—Permanent Virtual Circuit

ROSE—Remote Operations Service Element

SACF—Single Association Control Function

SAO—Single Association Object

SCAI—Switch-Computer Applications Interface

SCCP—Signaling Connection Control Part

SCP—Signaling Control Point

SDLC—Synchronous Data Link Control

SDU—Service Data Unit

SS7—Signaling System 7

STP—Signaling Transfer Point

SVC—Switched Virtual Service

T1AG—T1 Advisory Group

TCAP—Transaction Processing Application Part

TCP/IP—Transmission Control Protocol/Internet Protocol

TIA—Telecommunications Industry Association

USBS—User Signaling Bearer Service

VC—Virtual Call

WAN—Wide Area Network

Appendix

B

SCAI Protocol

This text is taken from the American National Standard for Telecommunications—Switch-Computer Applications Interface (SCAI) T1.626-1993.*

* Due to a printing error on the original edition of this standard, the year on the running head at the top of each page reads "ANSI T1.626-1992" instead of "ANSI T1.626-1993."

10 SCAI protocol

This protocol was derived directly from the text of the standard. If a discrepancy exists between the protocol and the text, the text shall take precedence.

10.1 Mapping between SCAI and ROSE

Table 11 maps SCAI FEs to a ROSE operation class and indicates the SCAI FE class.

Table 11 – Mapping FEs to ROSE operation classes

Functional Element	ROSE Operation Class	FE class
Initiate_Monitor	2	request
Cancel_Monitor	2	request
Change_Monitor_Filter	2	request
Query_Monitor	3	request - linked
Monitor_Report	3	information - linked
Make_Call	2	request
Answer_Call	2	request
Clear_Call	2	request
Predictive_Make_Call	2	request
Service_Initiated	5	event
Call_Originated	5	event
Call_Delivered	5	event
Call_Arrived	5	event
Call_Received	5	event
Call_Established	5	event
Call_Cleared	5	event
Call_Failed	5	event
Agent_Logged_On	5	event
Agent_Logged_Off	5	event
Agent_Ready	5	event
Agent_Not_Ready	5	event
Agent_Working_Not_Ready	5	event
Agent_Working_Ready	5	event
Conference_Existing_Calls	2	request
Drop_Conference_Party	2	request
Calls_Conferenced	5	event
Conference_Party_Dropped	5	event
Single_Step_Transfer	2	request
Consultation_Transfer	2	request
Call_Transferred	5	event
Hold_Call	2	request
Call_Held	5	event
Retrieve_Call	2	request
Call_Retrieved	5	event
Query_Feature	2	request
Set_Feature	2	request
Route_Request	3	request
Route_Selected	3	request
Routing_Toggle	2	request
Network_Reached	5	event
Call_Diverted	5	event

Table 12 maps the SCAI operations to the ROSE operations:

SCAI Operations	ROSE Operations
Request	Invoke
Success	Return Result
Error	Return Error

Table 12 – Mapping SCAI operations to ROSE operations

10.2 Module descriptions

The following OSI modules are identified for SCAI:

10.2.1 SCAI abstract-syntax-module-one

This module defines the remote operations and error definitions by type for SCAI. This includes the type definitions of all the data types (parameters) appearing in the operation and error type definitions.

10.2.2 SCAI abstract-syntax-module-two

This module defines abstract syntax specific to the SCAI protocol by allocating the operation and error code values. This module assigns unique (non-overlapping) values to the SCAI operations and SCAI errors required by the SCAI ase-module and SCAI security-ase-module. The type definitions for the operations and errors are imported from abstract-syntax-module-one. The application context modules then use the abstract syntax defined by this module.

10.2.3 SCAI ase-module

This module defines the services in the SCAI ASE in terms of consumer and performer invokes. It also identifies the consumer and performer (e.g., switch is the consumer). It imports the specific remote operations and their errors from abstract-syntax-module-two that make up the SCAI ASE. If at some later time a new grouping is found useful, a new module can be defined for that grouping.

10.2.4 SCAI security-ase-module

This module defines the type definitions, parameters, and services for the SCAI security ASE.

10.2.5 SCAI application-context-module-one

This module defines a particular application context for the association. This includes the ASEs (SCAI, ROSE, ACSE) and the abstract syntaxes of the data exchanged.

10.2.6 SCAI application-context-module-two

This module defines an application context for the association and contains all ASEs of application-context-module-one plus the SCAI security-ase-module.

10.2.7 Object Identifiers for SCAI – version one

0 ---> SCAI-abstract-syntax-module-one
1 ---> SCAI-ase-module
2 ---> SCAI-security-ase-module
3 ---> SCAI-application-context-module-one
4 ---> SCAI-application-context-module-two
5 ---> SCAI-abstract-syntax-module-two
6 ---> SCAI-ase
7 ---> SCAI-security-ase
8 ---> SCAI-context
9 ---> SCAI-security-context

{iso member-body (1)uSA (840) aNSI (1) sCAI(10021)
sCAI-abstract-syntax-module-one(0) version-one(0)}
{iso member-body (1)uSA (840) aNSI (1) sCAI(10021)
sCAI-ase-module(1) version-one(0)}
{iso member-body (1)uSA (840) aNSI (1) sCAI(10021)
sCAI-security-ase-module(2) version-one(0)}
{iso member-body (1)uSA (840) aNSI (1) sCAI(10021)
sCAI-application-context-module-one(3) version-one(0)}
{iso member-body (1)uSA (840) aNSI (1) sCAI(10021)
sCAI-application-context-module-two(4) version-one(0)}
{iso member-body (1)uSA (840) aNSI (1) sCAI(10021)
sCAI-abstract-syntax-module-two(5) version-one(0)}
{iso member-body (1)uSA (840) aNSI (1) sCAI(10021)
sCAI-ase(6) version-one(0)}
{iso member-body (1)uSA (840) aNSI (1) sCAI(10021)
sCAI-security-ase(7) version-one(0)}
{iso member-body (1)uSA (840) aNSI (1) sCAI(10021)
sCAI-context(8) version-one(0)}
{iso member-body (1)uSA (840) aNSI (1) sCAI(10021)
sCAI-security-context(9) version-one(0)}

10.3 ASN module definitions

10.3.1 SCAI abstract-syntax-module-one

```
SCAI-abstract-syntax-module-one
{iso member-body (1) uSA (840) aNSI (1) sCAI(10021)
                                        sCAI-abstract-syntax-module-one(0)version-one(0)}

DEFINITIONS ::=

BEGIN

EXPORTS                         -- Export of operation definitions

    AgentLoggedOff,
    AgentLoggedOn,
    AgentNotReady,
    AgentReady,
    AgentWorkingNotReady,
    AgentWorkingReady,
    AnswerCall,
    CallArrived,
    CallCleared,
    CallDelivered,
    CallDiverted,
    CallEstablished,
    CallFailed,
    CallHeld,
    CallOriginated,
    CallReceived,
    CallRetrieved,
    CallsConferenced,
    CallTransferred,
    CancelMonitor,
    ChangeMonitorFilter,
    ClearCall,
    ConferenceExistingCalls,
    ConferencePartyDropped,
    ConsultationTransfer,
    DropConferenceParty,
    HoldCall,
    InitiateMonitor,
    MakeCall,
    MonitorReport,
    NetworkReached,
    PredictiveMakeCall,
    QueryFeature,
    QueryMonitor,
    RetrieveCall,
    RouteRequest,
    RouteSelected,
    RoutingToggle,
    ServiceInitiated,
    SetFeature,
    SingleStepTransfer,
```

```
                              -- Exported Error Definitions
    AnswerCallError,
    CancelMonitorError,
    ChangeMonitorFilterError,
    ClearCallError,
    ConferenceExistingCallsError,
    ConsultationTransferError,
    DropConferencePartyError,
    GeneralError,
    HoldCallError,
    InitiateMonitorError,
    MakeCallError,
    MonitorReportError,
    PredictiveMakeCallError,
    PrivateError,
    QueryFeatureError,
    QueryMonitorError,
    RetrieveCallError,
    RouteRequestError,
    RouteSelectedError,
    RoutingToggleError,
    SetFeatureError,
    SingleStepTransferError;

IMPORTS          --      Imported application context and application service elements

    OPERATION, ERROR
    FROM  Remote-Operations-Notation
    {joint-iso-ccitt remote-operations(4) notation(0)};
```

10.3.1.1 Monitor functional group

10.3.1.1.1 InitiateMonitor

```
InitiateMonitor ::=              OPERATION
                                 ARGUMENT    InitiateMonitorArgument
                                 RESULT      InitiateMonitorResult
                                 ERRORS      {InitiateMonitorError, GeneralError, PrivateError}


InitiateMonitorArgument ::= SEQUENCE
  {
    device                                   DeviceID,
    requestedFilter                          FilterType                    OPTIONAL,
    privateData                              SCAIPrivateData               OPTIONAL}

InitiateMonitorResult ::= CHOICE
  {
    monitorID                                MonitorID,
    monitorIDplusPrivateData                 InitiateMonitorResultSequence}

InitiateMonitorResultSequence ::= SEQUENCE
  {
    monitorID                                MonitorID,
    privateData                              SCAIPrivateData}

InitiateMonitorError ::=         ERROR
                                 PARAMETER   InitiateMonitorErrorParameter

InitiateMonitorErrorParameter ::= SEQUENCE
  {
    errorNumber                              InitiateMonitorErrorNumber,
    defaultFilter          [0] IMPLICIT      FilterType                    OPTIONAL,
    errorInformation       [1] IMPLICIT      ErrorInformation              OPTIONAL}

InitiateMonitorErrorNumber ::= ENUMERATED
  {
    unknownDevice                            (0),
    invalidDevice                            (1),
    requestedFilterIsNull                    (2),
    invalidFilter                            (3),
    cannotOverrideDefaultFilter              (4),
    multipleFiltersNotSupported              (5),
    monitorLimit                             (6)}
```

10.3.1.1.2 CancelMonitor

```
CancelMonitor ::=                OPERATION
                                 ARGUMENT    CancelMonitorArgument
                                 RESULT      CancelMonitorResult
                                 ERRORS      {CancelMonitorError, GeneralError, PrivateError}


CancelMonitorArgument ::= CHOICE
  {
    monitorID                                MonitorID,
    monitorIDplusPrivateData                 CancelMonitorArgumentSequence}

CancelMonitorArgumentSequence ::= SEQUENCE
  {
    monitorID                                MonitorID,
    privateData                              SCAIPrivateData}

CancelMonitorResult ::= CHOICE
  {
    noData                                   NULL,
    privateData                              SCAIPrivateData}

CancelMonitorError ::=           ERROR
                                 PARAMETER   CancelMonitorErrorParameter

CancelMonitorErrorParameter ::= SEQUENCE
  {
    errorNumber                              CancelMonitorErrorNumber,
    errorInformation                         ErrorInformation              OPTIONAL}

CancelMonitorErrorNumber ::= ENUMERATED
  {
    unknownMonitorID                         (0),          -- the monitor ID is not known
    cannotCancelMonitor                      (1)}
```

10.3.1.1.3 ChangeMonitorFilter

```
ChangeMonitorFilter ::=          OPERATION
                                 ARGUMENT    ChangeMonitorFilterArgument
                                 RESULT      ChangeMonitorFilterResult
                                 ERRORS      {ChangeMonitorFilterError, GeneralError, PrivateError}

ChangeMonitorFilterArgument ::= SEQUENCE
  {
    monitorID                                MonitorID,
    newFilter                                FilterType,
    privateData                              SCAIPrivateData              OPTIONAL}

ChangeMonitorFilterResult ::= CHOICE
  {
    noData                                   NULL,
    privateData                              SCAIPrivateData}

ChangeMonitorFilterError ::=     ERROR
                                 PARAMETER   ChangeMonitorFilterErrorParameter

ChangeMonitorFilterErrorParameter ::= SEQUENCE
  {
    errorNumber                              ChangeMonitorFilterErrorNumber,
    defaultFilter         [0] IMPLICIT       FilterType                   OPTIONAL,
    errorInformation      [1] IMPLICIT       ErrorInformation             OPTIONAL}

ChangeMonitorFilterErrorNumber ::= ENUMERATED
  {
    unknownMonitorID                         (0),
    requestedFilterIsNull                    (1),
    invalidFilter                            (2),
    cannotOverrideDefaultFilter              (3)}
```

10.3.1.1.4 QueryMonitor

```
QueryMonitor ::=                OPERATION
                                ARGUMENT    QueryMonitorArgument
                                ERRORS      {QueryMonitorError, GeneralError, PrivateError}
                                LINKED      {QueryMonitor, MonitorReport}


QueryMonitorArgument ::= SEQUENCE
  {
    monitorID                               MonitorID                 OPTIONAL,
    privateData                             SCAIPrivateData           OPTIONAL}

QueryMonitorError ::=           ERROR
                                PARAMETER   QueryMonitorErrorParameter

QueryMonitorErrorParameter ::= SEQUENCE
  {
    errorNumber                             QueryMonitorErrorNumber,
    errorInformation                        ErrorInformation          OPTIONAL}

QueryMonitorErrorNumber ::= ENUMERATED
  {
    unknownMonitorID                        (0),
    cannotReportOnMonitor                   (1)}
```

10.3.1.1.5 MonitorReport

```
MonitorReport ::=               OPERATION
                                ARGUMENT     MonitorReportArgument
                                ERRORS       {MonitorReportError, GeneralError, PrivateError}


MonitorReportArgument ::= SEQUENCE
  {
    monitorID                                MonitorID,
    device                                   DeviceID,
    filter                                   FilterType,
    privateData                              SCAIPrivateData                 OPTIONAL}

MonitorReportError ::=          ERROR
                                PARAMETER    MonitorReportErrorParameter

MonitorReportErrorParameter ::= SEQUENCE
  {
    errorNumber                              MonitorReportErrorNumber,
    errorInformation                         ErrorInformation                OPTIONAL}

MonitorReportErrorNumber ::= NULL            -- None Defined
```

10.3.1.1.6 Monitor type references

```
FilterType ::= SEQUENCE
  {
    call              [0] IMPLICIT    CallFilter              DEFAULT {},
    callFeature       [1] IMPLICIT    CallServicesFilter      DEFAULT {},
    agent             [2] IMPLICIT    AgentFilter             DEFAULT {},
    conference        [3] IMPLICIT    ConferenceFilter        DEFAULT {}}

CallFilter ::= BIT STRING
  {
    serviceInitiated          (0),
    callOriginated            (1),
    callDelivered             (2),
    callEstablished           (3),
    callArrived               (4),
    callReceived              (5),
    callCleared               (6),
    callFailed                (7),
    networkReached            (8)}

AgentFilter ::= BIT STRING
  {
    agentLoggedOn             (0),
    agentLoggedOff            (1),
    agentReady                (2),
    agentNotReady             (3),
    agentWorkingNotReady      (4),
    agentWorkingReady         (5)}

CallServicesFilter ::= BIT STRING
  {
    callHeld                  (0),
    callRetrieved             (1),
    callTransfered            (2),
    divertEvent               (3)}

ConferenceFilter ::= BIT STRING
  {
    callsConferenced          (0),
    conferencePartyDropped    (1)}
```

10.3.1.2 Basic two-party call functional group

10.3.1.2.1 MakeCall

```
MakeCall ::=                OPERATION
                            ARGUMENT    MakeCallArgument
                            RESULT      MakeCallResult
                            ERRORS      {MakeCallError, GeneralError, PrivateError}

MakeCallArgument ::= SEQUENCE
  {
    callingDevice                       DeviceID,
    calledDevice                        DeviceID,
    callType                            CallType             DEFAULT voice,
    privateData                         SCAIPrivateData          OPTIONAL}

MakeCallResult ::= CHOICE
  {
    newCall                             CallID,
    newCallplusPrivateData              MakeCallResultSequence}

MakeCallResultSequence ::= SEQUENCE
  {
    newCall                             CallID,
    privateData                         SCAIPrivateData}

MakeCallError ::=           ERROR
                            PARAMETER   MakeCallErrorParameter

MakeCallErrorParameter ::= SEQUENCE
  {
    errorNumber                         MakeCallErrorNumber,
    errorInformation                    ErrorInformation         OPTIONAL}

MakeCallErrorNumber ::= ENUMERATED
  {
    unknownOriginatingDevice            (0),
    unknownDestinationDevice            (1),
    unknownCallType                     (2),
    invalidOriginatingDevice            (3),
    invalidCallType                     (4),
    callingDeviceOutOfService           (5)}
```

10.3.1.2.2 AnswerCall

```
AnswerCall ::=              OPERATION
                            ARGUMENT    AnswerCallArgument
                            RESULT      AnswerCallResult
                            ERRORS      {AnswerCallError, GeneralError, PrivateError}

AnswerCallArgument ::= SEQUENCE
  {
    partyToAnswer                       PartyID,
    privateData                         SCAIPrivateData                OPTIONAL}

AnswerCallResult ::= CHOICE
  {
    noData                              NULL,
    privateData                         SCAIPrivateData}

AnswerCallError ::=         ERROR
                            PARAMETER   AnswerCallErrorParameter

AnswerCallErrorParameter ::= SEQUENCE
  {
    errorNumber                         AnswerCallErrorNumber,
    errorInformation                    ErrorInformation               OPTIONAL}

AnswerCallErrorNumber ::= ENUMERATED
  {
    unknownAnsweringDevice              (0),
    unknownCallToAnswer                 (1),
    invalidAnsweringDevice              (2),
    invalidCallToAnswer                 (3),
    multipleCallsPresent                (4),
    unableToAnswer                      (5)}
```

10.3.1.2.3 ClearCall

```
ClearCall ::=                 OPERATION
                              ARGUMENT   ClearCallArgument
                              RESULT     ClearCallResult
                              ERRORS     {ClearCallError, GeneralError, PrivateError}


ClearCallArgument ::= SEQUENCE
 {
   partyToClear                          PartyID,
   privateData                           SCAIPrivateData                 OPTIONAL}

ClearCallResult ::= CHOICE
 {
   noData                                NULL,
   privateData                           SCAIPrivateData}

ClearCallError ::=            ERROR
                              PARAMETER  ClearCallErrorParameter

ClearCallErrorParameter ::= SEQUENCE
 {
   errorNumber                           ClearCallErrorNumber,
   errorInformation                      ErrorInformation                OPTIONAL}

ClearCallErrorNumber ::= ENUMERATED
 {
   unknownClearingDevice                 (0),
   unknownCallToClear                    (1),
   invalidClearingDevice                 (2),
   invalidCallToBeCleared                (3),
   noCallPresent                         (4),
   multipleCallsPresent                  (5)}
```

10.3.1.2.4 PredictiveMakeCall

```
PredictiveMakeCall ::=        OPERATION
                              ARGUMENT   PredictiveMakeCallArgument
                              RESULT     PredictiveMakeCallResult
                              ERRORS     {PredictiveMakeCallError, GeneralError, PrivateError}


PredictiveMakeCallArgument ::= SEQUENCE
  {
    callingDevice                          DeviceID,
    calledDevice                           DeviceID,
    callType              [0] IMPLICIT     CallType                     OPTIONAL,
    distribution          [1] IMPLICIT     DistributionType    DEFAULT  delivered,
    privateData                            SCAIPrivateData              OPTIONAL}

PredictiveMakeCallResult ::= CHOICE
  {
    newCall                                CallID,
    newCallplusPrivateData                 PredictiveMakeCallResultSequence}

PredictiveMakeCallResultSequence ::= SEQUENCE
  {
    newCall                                CallID,
    privateData                            SCAIPrivateData}

PredictiveMakeCallError ::=   ERROR
                              PARAMETER  PredicitiveMakeCallErrorParameter

PredictiveMakeCallErrorParameter ::= SEQUENCE
  {
    errorNumber                            PredictiveMakeCallErrorNumber,
    errorInformation                       ErrorInformation             OPTIONAL}

PredictiveMakeCallErrorNumber ::= ENUMERATED
  {
    unknownOriginatingDevice               (0),
    unknownDestinationDevice               (1),
    invalidOriginatingDevice               (2),
    invalidDestinationDevice               (3),
    invalidCallType                        (4),
    callingGroupOutOfService               (5)}

DistributionType ::= ENUMERATED
  {
    established                            (0),
    delivered                              (1)}
```

10.3.1.3 Basic two-party call progress functional group

10.3.1.3.1 ServiceInitiated

```
ServiceInitiated ::=            OPERATION
                                ARGUMENT    ServiceInitiatedArgument

ServiceInitiatedArgument ::= SEQUENCE
  {
    monitorID                               MonitorID,
    initiatedCall                           CallID,
    initiatingDevice                        DeviceID                        OPTIONAL,
    serviceInitiatedTransitionCause         ServiceInitiatedTransitionCause OPTIONAL,
    privateData                             SCAIPrivateData                 OPTIONAL}

ServiceInitiatedTransitionCause ::= ENUMERATED
  {
    serviceRequest                          (0)}
```

10.3.1.3.2 CallOriginated

```
CallOriginated ::=              OPERATION
                                ARGUMENT    CallOriginatedArgument

CallOriginatedArgument ::= SEQUENCE
  {
    monitorID                               MonitorID,
    originatedCall                          CallID,
    callingDevice           [1]             DeviceID                        OPTIONAL,
    calledDevice            [2]             DeviceID                        OPTIONAL,
    callOriginatedTransitionCause           CallOriginatedTransitionCause   OPTIONAL,
    privateData                             SCAIPrivateData                 OPTIONAL}

CallOriginatedTransitionCause ::= ENUMERATED
  {
    enBlocCallSetup                         (0),
    informationCollectionComplete           (1)}
```

10.3.1.3.3 CallDelivered

```
CallDelivered ::=              OPERATION
                               ARGUMENT    CallDeliveredArgument

CallDeliveredArgument ::= SEQUENCE
  {
    monitorID                             MonitorID,
    deliveredCall                         CallID,
    deliveredDevice          [4]          DeviceID                          OPTIONAL,
    callingDevice            [1]          DeviceID                          OPTIONAL,
    calledDevice             [2]          DeviceID                          OPTIONAL,
    callDeliveredTransitionCause          CallDeliveredTransitionCause      OPTIONAL,
    privateData                           SCAIPrivateData                   OPTIONAL}

CallDeliveredTransitionCause ::= ENUMERATED
  {
    destinationAlerting                   (0)}
```

10.3.1.3.4 CallArrived

```
CallArrived ::=                OPERATION
                               ARGUMENT    CallArrivedArgument

CallArrivedArgument ::= SEQUENCE
  {
    monitorID                             MonitorID,
    arrivedCall                           CallID,
    destinationDevice                     DeviceID,
    callingDevice            [1]          DeviceID                          OPTIONAL,
    calledDevice             [2]          DeviceID                          OPTIONAL,
    callArrivedTransitionCause            CallArrivedTransitionCause        OPTIONAL,
    privateData                           SCAIPrivateData                   OPTIONAL}

CallArrivedTransitionCause ::= ENUMERATED
  {
    newCallArrived                        (0),
    routeDetermined                       (1),
    reRouted                              (2),
    routingTimeout                        (3)}
```

10.3.1.3.5 CallReceived

```
CallReceived ::=                OPERATION
                                ARGUMENT   CallReceivedArgument

CallReceivedArgument ::= SEQUENCE
  {
    monitorID                                  MonitorID,
    receivedCall                               CallID,
    receivedDevice                             DeviceID,
    callingDevice              [1]             DeviceID                      OPTIONAL,
    calledDevice               [2]             DeviceID                      OPTIONAL,
    callReceivedTransitionCause                CallReceivedTransitionCause   OPTIONAL,
    privateData                                SCAIPrivateData               OPTIONAL}

CallReceivedTransitionCause ::= ENUMERATED
  {
    destinationAlerting                        (0)}
```

10.3.1.3.6 CallEstablished

```
CallEstablished ::=             OPERATION
                                ARGUMENT   CallEstablishedArgument

CallEstablishedArgument ::= SEQUENCE
  {
    monitorID                                  MonitorID,
    establishedCall                            CallID,
    view                                       EstablishedView,
    calledDevice                               DeviceID                         OPTIONAL,
    callEstablishedTransitionCause             CallEstablishedTransitionCause   OPTIONAL,
    privateData                                SCAIPrivateData                  OPTIONAL}

EstablishedView ::= CHOICE
  {
    originatingView            [0] IMPLICIT    EstablishedOriginatingView,
    terminatingView            [1] IMPLICIT    EstablishedTerminatingView}

EstablishedOriginatingView ::= SEQUENCE
  {
    callingDevice                              DeviceID,
    answeringDevice                            DeviceID                         OPTIONAL}

EstablishedTerminatingView ::= SEQUENCE
  {
    answeringDevice                            DeviceID,
    callingDevice                              DeviceID                         OPTIONAL}

CallEstablishedTransitionCause ::= ENUMERATED
  {
    destinationAnswered                        (0)}
```

10.3.1.3.7 CallCleared

```
CallCleared ::=                OPERATION
                               ARGUMENT    CallClearedArgument

CallClearedArgument ::= SEQUENCE
  {
    monitorID                            MonitorID,
    clearedCall                          CallID,
    clearingDevice                       DeviceID                    OPTIONAL,
    callClearedTransitionCause           CallClearedTransitionCause  OPTIONAL,
    privateData                          SCAIPrivateData             OPTIONAL}

CallClearedTransitionCause ::= ENUMERATED
  {
    callCleared                          (0),
    hostInitiatedClearing                (1),
    failureTreatmentComplete             (2)}
```

10.3.1.3.8 CallFailed

```
CallFailed ::=                  OPERATION
                                ARGUMENT     CallFailedArgument


CallFailedArgument ::= SEQUENCE
  {
    monitorID                               MonitorID,
    failedCall                              CallID,
    callingDevice              [1]          DeviceID                      OPTIONAL,
    calledDevice               [2]          DeviceID                      OPTIONAL,
    callFailedTransitionCause               CallFailedTransitionCause     OPTIONAL,
    privateData                             SCAIPrivateData               OPTIONAL}

CallFailedTransitionCause ::= ENUMERATED
  {
    unassignedDirectoryNumber               (0),
    noRouteToDestination                    (1),
    callRejected                            (2),
    busy                                    (3),
    numberChanged                           (4),
    networkOutOfOrder                       (5),
    temporaryFailure                        (6),
    switchingEquipmentCongestion            (7),
    requestedFacilityNotSubscribed          (8),
    serviceNotAvailable                     (9),
    noAnswer                                (10),
    authorizationFailure                    (11),
    informationCollectionTimeout            (12),
    destinationNotObtainable                (13),
    incompatibleDestination                 (15),
    invalidAccountCode                      (16),
    invalidInformationCollected             (17),
    informationAnalysisFailure              (18),
    destinationOutOfOrder                   (19),
    networkCongestion                       (20),
    networkNotObtainable                    (21),
    routingFailure                          (22),
    resourcesUnavailable                    (30),
    nocircuitAvailable                      (33),
    noUserResponding                        (34)}
```

10.3.1.4 Agent events functional group

10.3.1.4.1 AgentLoggedOn

```
AgentLoggedOn ::=            OPERATION
                             ARGUMENT     AgentLoggedOnArgument


AgentLoggedOnArgument ::=  SEQUENCE
  {
    monitorID                             MonitorID,
    agentDevice                           DeviceID,
    agent                 [0] IMPLICIT    AgentID                  OPTIONAL,
    agentGroup            [1]             AgentGroupID             OPTIONAL,
    privateData                           SCAIPrivateData          OPTIONAL}
```

10.3.1.4.2 AgentLoggedOff

```
AgentLoggedOff ::=           OPERATION
                             ARGUMENT     AgentLoggedOffArgument

AgentLoggedOffArgument ::=  SEQUENCE
  {
    monitorID                             MonitorID,
    agentDevice                           DeviceID,
    agent                 [0] IMPLICIT    AgentID                  OPTIONAL,
    agentGroup            [1]             AgentGroupID             OPTIONAL,
    privateData                           SCAIPrivateData          OPTIONAL}
```

10.3.1.4.3 AgentReady

```
AgentReady ::=               OPERATION
                             ARGUMENT     AgentReadyArgument

AgentReadyArgument ::= SEQUENCE
  {
    monitorID                             MonitorID,
    agentDevice                           DeviceID,
    agent                 [0] IMPLICIT    AgentID                  OPTIONAL,
    agentGroup            [1]             AgentGroupID             OPTIONAL,
    privateData                           SCAIPrivateData          OPTIONAL}
```

10.3.1.4.4 AgentNotReady

```
AgentNotReady ::=            OPERATION
                             ARGUMENT     AgentNotReadyArgument

AgentNotReadyArgument ::= SEQUENCE
  {
    monitorID                             MonitorID,
    agentDevice                           DeviceID,
    agent                 [0] IMPLICIT    AgentID                  OPTIONAL,
    agentGroup            [1]             AgentGroupID             OPTIONAL,
    privateData                           SCAIPrivateData          OPTIONAL}
```

10.3.1.4.5 AgentWorkingNotReady

```
AgentWorkingNotReady ::=        OPERATION
                                ARGUMENT    AgentWorkingNotReadyArgument

AgentWorkingNotReadyArgument ::= SEQUENCE
  {
    monitorID                               MonitorID,
    agentDevice                             DeviceID,
    agent             [0] IMPLICIT          AgentID             OPTIONAL,
    agentGroup        [1]                   AgentGroupID        OPTIONAL,
    privateData                             SCAIPrivateData     OPTIONAL}
```

10.3.1.4.6 AgentWorkingReady

```
AgentWorkingReady ::=           OPERATION
                                ARGUMENT    AgentWorkingReadyArgument

AgentWorkingReadyArgument ::= SEQUENCE
  {
    monitorID                               MonitorID,
    agentDevice                             DeviceID,
    agent             [0] IMPLICIT          AgentID             OPTIONAL,
    agentGroup        [1]                   AgentGroupID        OPTIONAL,
    privateData                             SCAIPrivateData     OPTIONAL}
```

10.3.1.5 Conference functional group

10.3.1.5.1 ConferenceExistingCalls

```
ConferenceExistingCalls ::= OPERATION
                            ARGUMENT    ConferenceExistingCallsArgument
                            RESULT      ConferenceExistingCallsResult
                            ERRORS      {ConferenceExistingCallsError, GeneralError, PrivateError}

ConferenceExistingCallsArgument ::= SEQUENCE
  {
    conferenceDevice                          DeviceID,
    call1               [0] IMPLICIT          CallID                  OPTIONAL,
    call2               [1] IMPLICIT          CallID                  OPTIONAL,
    privateData                               SCAIPrivateData         OPTIONAL}

ConferenceExistingCallsResult ::= CHOICE
  {
    conference                                CallID,
    conferenceIDplusParties                   ConferenceExistingCallsResultSequence}

ConferenceExistingCallsResultSequence ::= SEQUENCE
  {
    conference                                CallID,
    controller          [0]                   PartyID                 OPTIONAL,
    party2              [1]                   PartyID                 OPTIONAL,
    party3              [2]                   PartyID                 OPTIONAL,
    privateData                               SCAIPrivateData         OPTIONAL}

ConferenceExistingCallsError ::= ERROR
                              PARAMETER  ConferenceExistingCallsErrorParameter

ConferenceExistingCallsErrorParameter ::= SEQUENCE
  {
    errorNumber                               ConferenceExistingCallsErrorNumber,
    errorInformation                          ErrorInformation        OPTIONAL}

ConferenceExistingCallsErrorNumber ::= ENUMERATED
  {
  unknownConferencingDevice                   (0),
  unknownCall1                                (1),
  unknownCall2                                (2),
  invalidConferencingDevice                   (3),
  invalidCall1                                (4),
  invalidCall2                                (5),
  multipleCallsPresent                        (6)}
```

ANSI T1.626-1992

10.3.1.5.2 DropConferenceParty

```
DropConferenceParty ::=      OPERATION
                             ARGUMENT    DropConferencePartyArgument
                             RESULT      DropConferencePartyResult
                             ERRORS      {DropConferencePartyError, GeneralError, PrivateError}


DropConferencePartyArgument ::= SEQUENCE
  {
    conference                          CallID,
    partyToDrop                         PartyID,
    privateData                         SCAIPrivateData              OPTIONAL}

DropConferencePartyResult ::= CHOICE
  {
    noData                              NULL,
    privateData                         SCAIPrivateData}

DropConferencePartyError ::= ERROR
                             PARAMETER   DropConferencePartyErrorParameter

DropConferencePartyErrorParameter ::= SEQUENCE
  {
    errorNumber                         DropConferencePartyErrorNumber,
    errorInformation                    ErrorInformation             OPTIONAL}

DropConferencePartyErrorNumber ::= ENUMERATED
  {
    unknownConference                   (0),
    unknownCallIDForDropParty           (1),
    unknownDeviceIDForDropParty         (2),
    invalidParty                        (3),
    cannotDropParty                     (4)}
```

10.3.1.5.3 CallsConferenced

```
CallsConferenced ::=          OPERATION
                              ARGUMENT   CallsConferencedArgument

CallsConferencedArgument ::= SEQUENCE
  {
    monitorID                             MonitorID,
    conferencingDevice                    DeviceID,
    call1                                 CallID,
    call2                                 CallID,
    conference                            CallID,
    controller              [0]           PartyID              OPTIONAL,
    party2                  [1]           PartyID              OPTIONAL,
    party3                  [2]           PartyID              OPTIONAL,
    privateData                           SCAIPrivateData      OPTIONAL}
```

10.3.1.5.4 ConferencePartyDropped

```
ConferencePartyDropped ::= OPERATION
                              ARGUMENT   ConferencePartyDroppedArgument

ConferencePartyDroppedArgument ::= SEQUENCE
  {
    monitorID                             MonitorID,
    conference                            CallID,
    droppedParty                          PartyID,
    privateData                           SCAIPrivateData      OPTIONAL}
```

10.3.1.6 Transfer functional group

10.3.1.6.1 SingleStepTransfer

```
SingleStepTransfer ::=       OPERATION
                             ARGUMENT    SingleStepTransferArgument
                             RESULT      SingleStepTransferResult
                             ERRORS      {SingleStepTransferError, GeneralError, PrivateError}


SingleStepTransferArgument ::= SEQUENCE
  {
    transferringDevice                   DeviceID,
    destinationDevice                    DeviceID,
    establishedCall                      CallID                  OPTIONAL,
    privateData                          SCAIPrivateData         OPTIONAL}


SingleStepTransferResult ::= CHOICE
  {
    newCall                              CallID,
    newCallplusPrivateData               SingleStepTransferResultSequence}

SingleStepTransferResultSequence ::= SEQUENCE
  {
    newCall                              CallID                  OPTIONAL,
    privateData                          SCAIPrivateData         OPTIONAL}

SingleStepTransferError ::=  ERROR
                             PARAMETER   SingleStepTransferErrorParameter

SingleStepTransferErrorParameter ::= SEQUENCE
  {
    errorNumber                          SingleStepTransferErrorNumber,
    errorInformation                     ErrorInformation        OPTIONAL}

SingleStepTransferErrorNumber ::= ENUMERATED
  {
    unknownTransferringDevice            (0),
    unknownDestinationDevice             (1),
    unknownExistingCall                  (2),
    invalidTransferringDevice            (3),
    invalidDestinationDevice             (4),
    invalidExistingCall                  (5)}
```

10.3.1.6.2 ConsultationTransfer

```
ConsultationTransfer ::=        OPERATION
                                ARGUMENT   ConsultationTransferArgument
                                RESULT     ConsultationTransferResult
                                ERRORS     {ConsultationTransferError, GeneralError, PrivateError}

ConsultationTransferArgument ::= SEQUENCE
  {
    transferringDevice                     DeviceID,
    call1                  [0] IMPLICIT    CallID                      OPTIONAL,
    call2                  [1] IMPLICIT    CallID                      OPTIONAL,
    privateData                            SCAIPrivateData             OPTIONAL}

ConsultationTransferResult ::= CHOICE
  {
    newCall                                CallID,
    newCallplusPrivateData                 ConsultationTransferResultSequence}

ConsultationTransferResultSequence ::= SEQUENCE
  {
    newCall                                CallID                      OPTIONAL,
    privateData                            SCAIPrivateData             OPTIONAL}

ConsultationTransferError ::=   ERROR
                                PARAMETER  ConsultationTransferErrorParameter

ConsultationTransferErrorParameter ::= SEQUENCE
  {
    errorNumber                            ConsultationTransferErrorNumber,
    errorInformation                       ErrorInformation            OPTIONAL}

ConsultationTransferErrorNumber ::= ENUMERATED
  {
    unknownTransferringDevice              (0),
    unknownCall1                           (1),
    unknownCall2                           (2),
    invalidTransferringDevice              (3),
    invalidCall1                           (4),
    invalidCall2                           (5)}
```

10.3.1.6.3 CallTransferred

```
CallTransferred ::=          OPERATION
                             ARGUMENT     CallTransferredArgument


CallTransferredArgument ::= SEQUENCE
  {
    monitorID                                 MonitorID,
    transferringDevice                        DeviceID,
    calls                 [0]                 CallsInTransfer          OPTIONAL,
    newCall               [1] IMPLICIT        CallID                   OPTIONAL,
    destinationDevice     [2]                 DeviceID                 OPTIONAL,
    privateData                               SCAIPrivateData          OPTIONAL}

CallsInTransfer ::= CHOICE
  {
    singleStep                                CallID,
    consultation                              ConsultationTransferCalls}

ConsultationTransferCalls ::= SEQUENCE
  {
    call1                 [0] IMPLICIT        CallID                   OPTIONAL,
    call2                 [1] IMPLICIT        CallID                   OPTIONAL}
```

10.3.1.7 Hold-Retrieve functional group

10.3.1.7.1 HoldCall

```
HoldCall ::=                OPERATION
                            ARGUMENT    HoldCallArgument
                            RESULT      HoldCallResult
                            ERRORS      {HoldCallError, GeneralError, PrivateError}


HoldCallArgument ::= SEQUENCE
  {
    partyToHold                         PartyID,
    privateData                         SCAIPrivateData           OPTIONAL}

HoldCallResult ::= CHOICE
  {
    noData                              NULL,
    privateData                         SCAIPrivateData}

HoldCallError  ::=          ERROR
                            PARAMETER   HoldCallErrorParameter

HoldCallErrorParameter ::= SEQUENCE
  {
    errorNumber                         HoldCallErrorNumber,
    errorInformation                    ErrorInformation          OPTIONAL}

HoldCallErrorNumber ::= ENUMERATED
  {
    unknownHoldingDevice                (0),
    unknownActiveCall                   (1),
    invalidHoldingDevice                (2),
    invalidActiveCall                   (3),
    multipleCallsPresent                (4)}
```

10.3.1.7.2 CallHeld

```
CallHeld ::=                OPERATION
                            ARGUMENT    CallHeldArgument


CallHeldArgument ::= SEQUENCE
  {
    monitorID                           MonitorID,
    heldCall                            CallID,
    holdingDevice                       DeviceID,
    privateData                         SCAIPrivateData           OPTIONAL}
```

10.3.1.7.3 RetrieveCall

```
RetrieveCall ::=            OPERATION
                            ARGUMENT    RetrieveCallArgument
                            RESULT      RetrieveCallResult
                            ERRORS      {RetrieveCallError, GeneralError, PrivateError}


RetrieveCallArgument ::= SEQUENCE
  {
    partyToRetrieve                     PartyID,
    privateData                         SCAIPrivateData                   OPTIONAL}

RetrieveCallResult ::= CHOICE
  {
    noData                              NULL,
    privateData                         SCAIPrivateData}

RetrieveCallError ::=       ERROR
                            PARAMETER   RetrieveCallErrorParameter

RetrieveCallErrorParameter ::= SEQUENCE
  {
    errorNumber                         RetrieveCallErrorNumber,
    errorInformation                    ErrorInformation                  OPTIONAL}

RetrieveCallErrorNumber ::= ENUMERATED
  {
    unknownRetrievingDevice             (0),
    unknownHeldCall                     (1),
    invalidRetrievingDevice             (2),
    invalidHeldCall                     (3),
    multipleCallsPresent                (4)}
```

10.3.1.7.4 CallRetrieved

```
CallRetrieved ::=           OPERATION
                            ARGUMENT    CallRetrievedArgument


CallRetrievedArgument ::= SEQUENCE
  {
    monitorID                           MonitorID,
    retrievedCall                       CallID,
    retrievingDevice                    DeviceID,
    privateData                         SCAIPrivateData                   OPTIONAL}
```

10.3.1.8 Feature functional group

10.3.1.8.1 QueryFeature

```
QueryFeature ::=                OPERATION
                                ARGUMENT    QueryFeatureArgument
                                RESULT      QueryFeatureResult
                                ERRORS      {QueryFeatureError, GeneralError, PrivateError}

QueryFeatureArgument ::= SEQUENCE
  {
    device                                  DeviceID,
    feature                                 QueryDeviceFeature,
    privateData                             SCAIPrivateData                 OPTIONAL}

QueryFeatureResult ::= CHOICE
  {
    queryFeatureInfo                        QueryFeatureInformation,
    queryFeatureInfoplusPrivateData         QueryFeatureResultSequence}

QueryFeatureResultSequence ::= SEQUENCE
  {
    queryFeatureInfo                        QueryFeatureInformation,
    privateData                             SCAIPrivateData}

QueryFeatureError ::=           ERROR
                                PARAMETER   QueryFeatureErrorParameter

QueryFeatureErrorParameter ::= SEQUENCE
  {
    errorNumber                             QueryFeatureErrorNumber,
    errorInformation                        ErrorInformation                OPTIONAL}

QueryFeatureErrorNumber ::= ENUMERATED
  {
    unknownDevice                           (0),
    unknownFeature                          (1),
    invalidDevice                           (2),
    invalidFeature                          (3)}

QueryDeviceFeature ::= CHOICE               -- additional choices may be identified in the future
  {
    agent                   [5]             QueryAgentFeature}

QueryAgentFeature ::= CHOICE
  {
    group                                   AgentGroupID,
    groupNotSupplied                        NULL}

QueryFeatureInformation ::= CHOICE          -- additional choices may be identified in the future
  {
    queryAgentFeatureInfo  [5]  IMPLICIT    QueryAgentFeatureInformation}

QueryAgentFeatureInformation ::= SEQUENCE
  {
    agentState                              AgentState,
    agent                   [0] IMPLICIT    AgentID                         OPTIONAL,
    agentGroup              [1]             AgentGroupID                    OPTIONAL}
```

10.3.1.8.2 SetFeature

```
SetFeature ::=                OPERATION
                              ARGUMENT     SetFeatureArgument
                              RESULT       SetFeatureResult
                              ERRORS       {SetFeatureError, GeneralError, PrivateError}

SetFeatureArgument ::= SEQUENCE
 {
   device                                  DeviceID,
   setFeatureInfo                          SetFeatureInformation,
   privateData                             SCAIPrivateData                  OPTIONAL}

SetFeatureResult ::= CHOICE
 {
   noData                                  NULL,
   privateData                             SCAIPrivateData}

SetFeatureError ::=           ERROR
                              PARAMETER    SetFeatureErrorParameter

SetFeatureErrorParameter ::= SEQUENCE
 {
   errorNumber                             SetFeatureErrorNumber,
   errorInformation                        ErrorInformation                 OPTIONAL}

SetFeatureErrorNumber ::= ENUMERATED
 {
   unknownDevice                           (0),
   unknownFeature                          (1),
   invalidFeature                          (2),
   unknownAgentIdentifier                  (3),
   invalidAgentIdentifier                  (4),
   invalidAgentPassword                    (5),
   unknownAgentGroupIdentifier             (6),
   invalidAgentGroupIdentifier             (7)}

SetFeatureInformation ::= CHOICE
 {
   agentFeature           [3]              SetAgentFeature}

SetAgentFeature ::= CHOICE
 {
   loggedOn               [0] IMPLICIT     LogOnInfo,
   loggedOff              [1] IMPLICIT     AgentInfo,
   notReady               [2] IMPLICIT     AgentInfo,
   ready                  [3] IMPLICIT     AgentInfo,
   workingNotReady        [4] IMPLICIT     AgentInfo,
   workingReady           [5] IMPLICIT     AgentInfo}

LogOnInfo ::= SEQUENCE
 {
   agent                  [0] IMPLICIT     AgentID                          OPTIONAL,
   group                  [1]              AgentGroupID                     OPTIONAL,
   password               [2] IMPLICIT     AgentPassword                    OPTIONAL}
```

```
AgentInfo ::= SEQUENCE
  {
    agent              [0] IMPLICIT     AgentID                          OPTIONAL,
    group              [1]              AgentGroupID                     OPTIONAL}
```

10.3.1.8.3 Agent feature type references

```
AgentID ::= OCTET STRING

AgentGroupID ::= CHOICE
  {
    groupID            [0] IMPLICIT     OCTET STRING,
    deviceID           [1]              DeviceID}

AgentPassword ::= OCTET STRING

AgentState ::= ENUMERATED
  {
    null                                (0),
    ready                               (1),
    notReady                            (2),
    workingReady                        (3),
    workingNotReady                     (4)}
```

10.3.1.9 Routing functional group

10.3.1.9.1 RouteRequest

```
RouteRequest ::=            OPERATION
                            ARGUMENT     RouteRequestArgument
                            ERRORS       {RouteRequestError, GeneralError, PrivateError}


RouteRequestArgument ::= SEQUENCE
  {
    callToRoute                          CallID,
    destinationDevice                    DeviceID,
    callingDevice          [1]           DeviceID                OPTIONAL,
    calledDevice           [2]           DeviceID                OPTIONAL,
    privateData                          SCAIPrivateData         OPTIONAL}

RouteRequestError ::=       ERROR
                            PARAMETER  RouteRequestErrorParameter

RouteRequestErrorParameter ::= SEQUENCE
  {
    errorNumber                          RouteRequestErrorNumber,
    errorInformation                     ErrorInformation        OPTIONAL}

RouteRequestErrorNumber ::= NULL         -- None Defined
```

10.3.1.9.2 RouteSelected

```
RouteSelected ::=          OPERATION
                           ARGUMENT   RouteSelectedArgument
                           ERRORS     {RouteSelectedError, GeneralError, PrivateError}

RouteSelectedArgument ::= SEQUENCE
  {
    callToRoute                        CallID,
    alternateDestinationDevice         DeviceID              OPTIONAL,
    privateData                        SCAIPrivateData       OPTIONAL}

RouteSelectedError  ::=    ERROR
                           PARAMETER  RouteSelectedErrorParameter

RouteSelectedErrorParameter ::= SEQUENCE
  {
    errorNumber                        RouteSelectedErrorNumber,
    errorInformation                   ErrorInformation      OPTIONAL}

RouteSelectedErrorNumber ::= ENUMERATED
  {
    unknownCallToRoute                 (0),
    unknownDestinationDevice           (1),
    invalidCallToRoute                 (2),
    invalidDestinationDevice           (3)}
```

10.3.1.9.3 RoutingToggle

```
RoutingToggle ::=              OPERATION
                               ARGUMENT    RoutingToggleArgument
                               ERRORS      {RoutingToggleError, GeneralError, PrivateError}

RoutingToggleArgument ::= SEQUENCE
  {
    device                                 DeviceID ,
    routingActivate                        BOOLEAN,
    privateData                            SCAIPrivateData               OPTIONAL}

RoutingToggleError ::=         ERROR
                               PARAMETER   RoutingToggleErrorParameter

RoutingToggleErrorParameter ::= SEQUENCE
  {
    errorNumber                            RoutingToggleErrorNumber,
    errorInformation                       ErrorInformation              OPTIONAL}

RoutingToggleErrorNumber ::= ENUMERATED
  {
    unknownDevice                          (0),
    invalidDevice                          (1),
    cannotDisableRouting                   (2),
    cannotEnableRouting                    (3)}
```

10.3.1.10 Call diversion functional group

10.3.1.10.1 CallDiverted

```
CallDiverted ::=              OPERATION
                              ARGUMENT    CallDivertedArgument


CallDivertedArgument ::= SEQUENCE
  {
    monitorID                             MonitorID,
    divertedCall                          CallID,
    divertingDevice                       DeviceID,
    newDestination                        DeviceID               OPTIONAL,
    diversionCause                        DiversionCause         OPTIONAL,
    privateData                           SCAIPrivateData        OPTIONAL}

DiversionCause ::= ENUMERATED
  {
    forwardNoAnswer                       (0),
    forwardBusy                           (1),
    forwardUnconditional                  (2)}
```

10.3.1.11 Other functional group

10.3.1.11.1 NetworkReached

```
NetworkReached ::=            OPERATION
                              ARGUMENT    NetworkReachedArgument


NetworkReachedArgument ::= SEQUENCE
  {
    monitorID                             MonitorID,
    existingCall                          CallID,
    callingDevice         [1]             DeviceID               OPTIONAL,
    calledDevice          [2]             DeviceID               OPTIONAL,
    privateData                           SCAIPrivateData        OPTIONAL}
```

10.3.1.12 Common type references

```
CallID ::= OCTET STRING

PartyID ::= CHOICE
  {
    callID                [0] IMPLICIT    CallID,
    deviceID              [1]             DeviceID,
    callIDplusDeviceID                    CallIDplusDeviceID}

CallIDplusDeviceID ::= SEQUENCE
  {
    callID                                CallID,
    deviceID                              DeviceID}

CallType ::= ENUMERATED
  {
    voice                   (0),
    data                    (1)}

SCAIPrivateData ::= EXTERNAL

MonitorID ::= CHOICE
  {
    monitorID                             OCTET STRING,
    default                               NULL}

-- Device ID definition starts here

DeviceID ::= CHOICE
  {
    dialingNumber                         NumberDigits,
    deviceNumber                          DeviceNumber,
    explicitNumber                        ExpandedTON,
    qualifiedNumber                       QualifiedNumber,
    dynamicID                             DynamicID}

ExpandedTON ::= CHOICE
  {
    explicitPublic        [3]             PublicTON,
    explicitPrivate       [5]             PrivateTON}

QualifiedNumber ::= SEQUENCE
  {
    number                                NumberType,
    numberPlan            [0] IMPLICIT    NumberPlan              OPTIONAL,
    presentation          [1] IMPLICIT    PresentationIndicator   OPTIONAL,
    screening             [2] IMPLICIT    ScreeningIndicator      OPTIONAL,
    subAddress                            Subaddress              OPTIONAL}

NumberType ::= CHOICE
  {
    dialingNumber                         NumberDigits,
    deviceNumber                          DeviceNumber,
    explicitNumber                        ExpandedTON}
```

```
PublicTON ::= CHOICE
  {
    unknown             [0] IMPLICIT      IA5String,
    international       [1] IMPLICIT      IA5String,
    national            [2] IMPLICIT      IA5String,
    networkspecific     [3] IMPLICIT      IA5String,
    subscriber          [4] IMPLICIT      IA5String,
    abbreviated         [5] IMPLICIT      IA5String}

PrivateTON ::= CHOICE
  {
    unknown             [0] IMPLICIT      IA5String,
    level3              [1] IMPLICIT      IA5String,
    level2              [2] IMPLICIT      IA5String,
    level1              [3] IMPLICIT      IA5String,
    networkspecific     [4] IMPLICIT      IA5String,
    localNumber         [5] IMPLICIT      IA5String,
    abbreviated         [6] IMPLICIT      IA5String}

NumberPlan ::= ENUMERATED
  {
    unknown                               (0),        -- No knowledge
    e164                                  (1),        -- ISDN Numbering
    x121                                  (2),        -- data Numbering
    f69                                   (3),        -- telex numbering
    national                              (4),        -- national standard
    private                               (5)}        -- private numbering

PresentationIndicator ::= ENUMERATED
  {
    allowed                               (0),
    restricted                            (1),
    notAvailable                          (2)}

ScreeningIndicator ::= ENUMERATED
  {
    upns                                  (0),        -- User-provided, not screened
    upvp                                  (1),        -- User-provided, screened and passed
    upvf                                  (2),        -- User-provided, screened and failed
    networkprovided                       (3)}

SubAddress ::= SEQUENCE
  {
    typeOfSubAddress                      TypeOfSubAddress,
    subAddressInformation                 OCTET STRING}

TypeOfSubAddress ::= ENUMERATED
  {
    nSAP                                  (0),
    userSpecified                         (1)}

NumberDigits ::= IA5String

DeviceNumber ::= INTEGER

DynamicID ::= OCTET STRING
```

10.3.1.13 General errors

```
GeneralError ::=              ERROR
                              PARAMETER   GeneralErrorParameter

GeneralErrorParameter ::= SEQUENCE
  {
    errorNumber                           GeneralErrorNumber,
    errorInformation                      ErrorInformation          OPTIONAL}

GeneralErrorNumber ::= ENUMERATED
  {
    privilegeError                        (0),
    resourceNotAvailable                  (1),
    otherError                            (2)}

ErrorInformation ::= SEQUENCE
  {
    parameterInError                      ParameterInError          OPTIONAL,
    privateData                           SCAIPrivateData           OPTIONAL}

ParameterInError ::= CHOICE
  {
    callID              [0] IMPLICIT      CallID,
    deviceID            [1]               DeviceID,
    callType            [2] IMPLICIT      CallType,
    partyID             [3]               PartyID,
    monitorID           [4]               MonitorID,
    monitorFilter       [5] IMPLICIT      FilterType,
    agentID             [6] IMPLICIT      AgentID,
    agentGroupID        [7]               AgentGroupID,
    agentPassword       [8] IMPLICIT      AgentPassword}

PrivateError ::=              ERROR
                              PARAMETER   PrivateErrorParameter

PrivateErrorParameter ::= EXTERNAL
```

10.3.1.14 Abort Information

```
AbortInfo ::= SEQUENCE
  {
    requestor                          AbortRequestor,
    privateData                        SCAIPrivateData                 OPTIONAL}

AbortRequestor ::= CHOICE
     {
       sCAIServiceProvider  [0] IMPLICIT   SCAIServiceProviderAbortCause,
       sCAISecurityProvider [1] IMPLICIT   SCAISecurityProviderAbortCause,
       sCAIServiceUser      [2] IMPLICIT   SCAIServiceUserAbortCause}

SCAIServiceProviderAbortCause ::= ENUMERATED
  {
    sCAI-error                  (0),
    security-error              (1)}

SCAISecurityProviderAbortCause ::= ENUMERATED
  {
    unspecified                 (0),
    sequenceNumberViolation     (1),
    timeStampViolation          (2),
    senderIDViolation           (3),
    integrityViolation          (4),
    confidentialityError        (5)}

SCAIServiceUserAbortCause ::= ENUMERATED
  {
    unspecified                 (0)}


END          -- end of SCAI abstract-syntax-module-one
```

ANSI T1.626-1992

10.3.2 SCAI abstract-syntax-module-two

```
SCAI-abstract-syntax-module-two
{iso member-body (1) uSA (840) aNSI (1) sCAI(10021)
                                        sCAI-abstract-syntax-module-two(5)version-one(0)}

DEFINITIONS ::=

BEGIN

EXPORTS

-- operation values

agentLoggedOff              AgentLoggedOff              ::=  0
agentLoggedOn               AgentLoggedOn               ::=  1
agentNotReady               AgentNotReady               ::=  2
agentReady                  AgentReady                  ::=  3
agentWorkingNotReady        AgentWorkingNotReady        ::=  4
agentWorkingReady           AgentWorkingReady           ::=  5
answerCall                  AnswerCall                  ::=  6
singleStepTransfer          SingleStepTransfer          ::=  7
callArrived                 CallArrived                 ::=  8
callCleared                 CallCleared                 ::=  9
callsConferenced            CallsConferenced            ::=  10
callDelivered               CallDelivered               ::=  11
callDiverted                CallDiverted                ::=  12
callEstablished             CallEstablished             ::=  13
callFailed                  CallFailed                  ::=  14
callHeld                    CallHeld                    ::=  15
callOriginated              CallOriginated              ::=  16
callReceived                CallReceived                ::=  17
callRetrieved               CallRetrieved               ::=  18
callTransferred             CallTransferred             ::=  20
cancelMonitor               CancelMonitor               ::=  21
changeMonitorFilter         ChangeMonitorFilter         ::=  22
clearCall                   ClearCall                   ::=  23
conferenceExistingCalls     ConferenceExistingCalls     ::=  24
conferencePartyDropped      ConferencePartyDropped      ::=  25
consultationTransfer        ConsultationTransfer        ::=  26
dropConferenceParty         DropConferenceParty         ::=  27
holdCall                    HoldCall                    ::=  28
initiateMonitor             InitiateMonitor             ::=  29
makeCall                    MakeCall                    ::=  30
monitorReport               MonitorReport               ::=  31
networkReached              NetworkReached              ::=  32
predictiveMakeCall          PredictiveMakeCall          ::=  33
queryFeature                QueryFeature                ::=  34
queryMonitor                QueryMonitor                ::=  35
retrieveCall                RetrieveCall                ::=  36
routeRequest                RouteRequest                ::=  37
routeSelected               RouteSelected               ::=  38
routingToggle               RoutingToggle               ::=  39
serviceInitiated            ServiceInitiated            ::=  40
setFeature                  SetFeature                  ::=  41
```

```
-- security operation values

ss-transfer                         SS-Transfer                         ::=   42

-- error values

answerCallError                     AnswerCallError                     ::=   0
singleStepTransferError             SingleStepTransferError             ::=   1
cancelMonitorError                  CancelMonitorError                  ::=   2
changeMonitorFilterError            ChangeMonitorFilterError            ::=   3
clearCallError                      ClearCallError                      ::=   4
conferenceExistingCallsError        ConferenceExistingCallsError        ::=   5
consultationTransferError           ConsultationTransferError           ::=   7
dropConferencePartyError            DropConferencePartyError            ::=   8
generalError                        GeneralError                        ::=   9
holdCallError                       HoldCallError                       ::=   10
initiateMonitorError                InitiateMonitorError                ::=   11
makeCallError                       MakeCallError                       ::=   12
monitorReportError                  MonitorReportError                  ::=   13
predictiveMakeCallError             PredictiveMakeCallError             ::=   14
privateError                        PrivateError                        ::=   15
queryFeatureError                   QueryFeatureError                   ::=   16
queryMonitorError                   QueryMonitorError                   ::=   17
retrieveCallError                   RetrieveCallError                   ::=   18
routeRequestError                   RouteRequestError                   ::=   19
routeSelectedError                  RouteSelectedError                  ::=   20
routingToggleError                  RoutingToggleError                  ::=   21
setFeatureError                     SetFeatureError                     ::=   22;

IMPORTS

    AgentLoggedOff,                 -- operation types
    AgentLoggedOn,
    AgentNotReady,
    AgentReady,
    AgentWorkingNotReady,
    AgentWorkingReady,
    AnswerCall,
    CallArrived,
    CallCleared,
    CallsConferenced,
    CallDelivered,
    CallDiverted,
    CallEstablished,
    CallFailed,
    CallHeld,
    CallOriginated,
    CallReceived,
    CallRetrieved,
    CallTransferred,
    CancelMonitor,
    ChangeMonitorFilter,
    ClearCall,
    ConferenceExistingCalls,
    ConferencePartyDropped,
    ConsultationTransfer,
    DropConferenceParty,
    HoldCall,
```

```
      InitiateMonitor,
      MakeCall,
      MonitorReport,
      NetworkReached,
      PredictiveMakeCall,
      QueryFeature,
      QueryMonitor,
      RetrieveCall,
      RouteRequest,
      RouteSelected,
      RoutingToggle,
      ServiceInitiated,
      SetFeature,
      SingleStepTransfer,
                                        -- error types
      AnswerCallError,
      CancelMonitorError,
      ChangeMonitorFilterError,
      ClearCallError,
      ConferenceExistingCallsError,
      ConsultationTransferError,
      DropConferencePartyError,
      GeneralError,
      HoldCallError,
      InitiateMonitorError,
      MakeCallError,
      MonitorReportError ,
      PredictiveMakeCallError,
      PrivateError,
      QueryFeatureError,
      QueryMonitorError,
      RetrieveCallError,
      RouteRequestError,
      RouteSelectedError,
      RoutingToggleError,
      SetFeatureError,
      SingleStepTransferError

FROM SCAI-abstract-syntax -module-one
{iso member-body (1) uSA (840) aNSI (1) sCAI(10021)
                                        sCAI-abstract-syntax-module-one(0)version-one(0)}

SS-Transfer
FROM SCAI-security-ase-module
{iso member-body (1) uSA (840) aNSI (1) sCAI(10021) sCAI-security-ase-module(2)version-one(0)};

END            -- SCAI abstract-syntax-module-two
```

10.3.3 SCAI ase-module

```
SCAI-ase-module
{iso  member-body (1) uSA (840) aNSI (1) sCAI(10021)
                                        sCAI-ase-module(1)version-one(0)}
DEFINITIONS ::=

BEGIN

EXPORTS  sCAI-ase;

IMPORTS
    agentLoggedOff,
    agentLoggedOn,
    agentNotReady,
    agentReady,
    agentWorkingNotReady,
    agentWorkingReady,
    answerCall,
    callArrived,
    callCleared,
    callDelivered,
    callDiverted,
    callEstablished,
    callFailed,
    callHeld,
    callOriginated,
    callReceived,
    callRetrieved,
    callsConferenced,
    callTransferred,
    cancelMonitor,
    changeMonitorFilter,
    clearCall,
    conferenceExistingCalls,
    conferencePartyDropped,
    consultationTransfer,
    dropConferenceParty,
    holdCall,
    initiateMonitor,
    makeCall,
    monitorReport,
    networkReached,
    predictiveMakeCall,
    queryFeature,
    queryMonitor,
    retrieveCall,
    routeRequest,
    routeSelected,
    routingToggle,
    serviceInitiated,
    setFeature,
    singleStepTransfer

FROM  SCAI-abstract-syntax-module-two
{iso  member-body (1) uSA (840) aNSI (1) sCAI(10021)
                                        sCAI-abstract-syntax-module-two(5)version-one(0)}
```

ANSI T1.626-1992

```
APPLICATION-SERVICE-ELEMENT
FROM  Remote-Operations-Notation-extension
{ joint-iso-ccitt remote operations (4) notation-extensions(2)};

SCAI-ase ::=  APPLICATION-SERVICE-ELEMENT

CONSUMER INVOKES                                -- Switch invokes
  {
     agentLoggedOff,
     agentLoggedOn,
     agentNotReady,
     agentReady,
     agentWorkingNotReady,
     agentWorkingReady,
     callArrived,
     callCleared,
     callDelivered,
     callDiverted,
     callEstablished,
     callFailed,
     callHeld,
     callOriginated,
     callReceived,
     callRetrieved,
     callsConferenced,
     callTransferred,
     cancelMonitor,
     conferencePartyDropped,
     monitorReport,
     networkReached,
     routeRequest,
     serviceInitiated}

SUPPLIER INVOKES                                -- computer invokes
  {
     answerCall,
     singleStepTransfer,
     cancelMonitor,
     changeMonitorFilter,
     clearCall,
     conferenceExistingCalls,
     consultationTransfer,
     dropConferenceParty,
     holdCall,
     initiateMonitor,
     makeCall,
     predictiveMakeCall,
     queryFeature,
     queryMonitor,
     retrieveCall,
     routeSelected,
     routingToggle,
     setFeature}

sCAI-ase  SCAI-ase  ::=
{iso  member-body (1) uSA (840) aNSI (1) sCAI(10021) sCAI-ase(6)version-one(0)}

END              -- end of SCAI-ase-module
```

10.3.4 SCAI security-ase-module

```
SCAI-security-ase-module
{iso  member-body (1) uSA (840) aNSI (1) sCAI(10021)
                                              sCAI-security-ase-module(2)version-one(0)}

DEFINITIONS ::=

BEGIN

EXPORTS sCAI-security-ase, SS-Transfer;

IMPORTS

APPLICATION-SERVICE-ELEMENT
FROM Remote-Operation-Notation
{joint-iso-ccitt remote-operations(4) notation-extention(2)}

DistingusihedName
FROM InformationFramework
{joint-iso-ccitt ds(5) modules(1) informationFramework(1)}

AlgorithmIdentifier, SIGNED, ENCRYPTED
FROM AuthenticationFramework
{joint-iso-ccitt de(5) modules(1) authentication-framework(7)}

ss-transfer
FROM   SCAI-abstract-syntax-module-two
{iso  member-body (1) uSA (840) aNSI (1) sCAI(10021)
                                              sCAI-abstract-syntax-module-two(5)version-one(0)};

SCAI-security-ase ::=  APPLICATION-SERVICE-ELEMENT

  OPERATIONS {ss-transfer}

sCAI-security-ase  SCAI-security-ase  ::=
{iso  member-body (1) uSA (840) aNSI (1) sCAI(10021) sCAI-security-ase(7)version-one(0)}

SS-Transfer ::=  SIGNED SEQUENCE
  {
    roAPDU                              RO-APDU,
    sender-ID                           DistinguishedName          OPTIONAL,
    confidentiality-algorithm-ID        AlgorithmIdentifier        OPTIONAL,
    message-sequence-number             INTEGER                    OPTIONAL,
    time-stamp                          UTCTime                    OPTIONAL}

RO-APDU ::=  CHOICE
  {
    sCAI-operation                      ROSEapdus,
    confidential-sCAI-operation  ENCRYPTED  ROSEapdus}

    --   for a description of ROSEapdus see Remote-Operations-APDUs
    --   {joint-iso-ccitt remote-operations(4) apdus(1)}

END            -- SCAI security-ase-module
```

10.3.5 SCAI application-context-module-one

```
SCAI-application-context-module-one
{iso member-body (1) uSA (840) aNSI (1) sCAI(10021)
                                        sCAI-application-context-module-one(3)version-one(0)}

DEFINITIONS ::=

BEGIN

EXPORTS  sCAI-context;

IMPORTS

APPLICATION-CONTEXT, aCSE
FROM Remote-Operations-Notation-extension
{ joint-iso-ccitt remote operations (4) notation-extensions(2)}

rOSE
FROM  Remote-Operations-APDUs
{joint-iso-ccitt remote-operations(4) apdus(1)}

sCAI-ase
FROM  SCAI-ase-module
{iso member-body (1) uSA (840) aNSI (1) sCAI(10021) sCAI-ase-module(1)version-one(0)};

SCAI-context  ::=  APPLICATION-CONTEXT

APPLICATIONS SERVICE ELEMENTS           {aCSE}

REMOTE OPERATIONS                       {rOSE}

  OPERATIONS OF                         {sCAI-ase}

ABSTRACT SYNTAXES
  {  {joint-iso-ccitt association-control (2) abstract-syntax(1) apdus(0)version(1)},
     {joint-iso-ccitt remote-operations(4) apdus(1)},
     {iso  member-body (1) uSA (840) aNSI (1) sCAI(10021)
                                        sCAI-abstract-syntax-module-one(0)version-one(0)},
     {iso  member-body (1) uSA (840) aNSI (1) sCAI(10021)
                                        sCAI-abstract-syntax-module-two(5)version-one(0)}};

sCAI-context  SCAI-context ::=
{iso  member-body (1) uSA (840) aNSI (1) sCAI(10021) sCAI-context(8)version-one(0)}

END             -- SCAI application-context-module-one
```

10.3.6 SCAI application-context-module-two

```
SCAI-application-context-module-two
{iso member-body (1) uSA (840) aNSI (1) sCAI(10021)
                                        sCAI-application-context-module-two(4)version-one(0)}

DEFINITIONS ::=

BEGIN

EXPORTS  sCAI-security-context;

IMPORTS

APPLICATION-CONTEXT, aCSE
FROM Remote-Operations-Notation-extension
{ joint-iso-ccitt remote-operations (4) notation-extensions(2)}

rOSE
FROM  Remote-Operations-APDUs
{joint-iso-ccitt remote-operations(4) apdus(1)}

sCAI-ase
FROM  SCAI-ase-module
{iso  member-body (1) uSA (840) aNSI (1) sCAI(10021) sCAI-ase-module(1)version-one(0)}

sCAI-security-ase-module
FROM  SCAI-security-ase-module
{iso  member-body (1) uSA (840) aNSI (1) sCAI(10021) sCAI-security-ase-module(2)version-one(0)};

SCAI-security-context   ::=   APPLICATION-CONTEXT

APPLICATIONS SERVICE ELEMENTS           {aCSE}

REMOTE OPERATIONS                       {rOSE}

OPERATIONS OF                           {sCAI-ase, sCAI-security-ase}

ABSTRACT SYNTAXES
{  {joint-iso-ccitt association-control (2) abstract-syntax(1) apdus(0) version(1)},
   {joint-iso-ccitt remote-operations(4) apdus(1)},
   {iso  member-body (1) uSA (840) aNSI (1) sCAI(10021)
                                        sCAI-abstract-syntax-module-one(0)version-one(0)},
   {iso  member-body (1) uSA (840) aNSI (1) sCAI(10021)
                                        sCAI-abstract-syntax-module-two(5)version-one(0)},
   {iso  member-body (1) uSA (840) aNSI (1) sCAI(10021)
                                        sCAI-security-ase-module(2)version-one(0)}};

sCAI-security-context  SCAI-security-context ::=
{iso  member-body (1) uSA (840) aNSI (1) sCAI(10021)
                                        sCAI-security-context(9)version-one(0)}

END             -- SCAI application-context-module-two
```

Bibliography

ANSI T1.114-1992; Signaling System Number 7 (SS7)—Transaction Capabilities Application Part (TCAP)

ANSI T1.606-1990; ISDN—Architectural Framework and Service Description for Frame-Relaying Bearer Service

ANSI T1.607-1990; ISDN—Layer 3 Signaling Specification for Circuit-Switched Bearer Service for DSS1.

ANSI T1.608-1991; ISDN—Signaling Specification for X.25 Packet-Switched Bearer Service for ISDN.

ANSI T1.615-1992; DSS1—Layer 3 Overview.

ANSI T1.617-1991; ISDN—Signaling Specification for Frame-Relay Bearer Service for DSS1.

ANSI T1.618-1991; ISDN—Core Aspects of Frame Relay Protocol for Use with Frame Relay Bearer Service.

ANSI T1.621-1992; ISDN—User-to-User Signaling Supplementary Service.

ANSI T1.623-1993; DSS1—Signaling Specification for the User Signaling Bearer Service.

ANSI T1.626-1993; Switch-Computer Applications Interface (SCAI).

ITU-T Q.701; Functional Description of the Message Transfer Part (MTP) of Signaling System No. 7.

ITU-T Q.705; Signaling Network Structure.

ITU-T Q.711; Functional Description of Signaling Connection Control Part.

ITU-T Q.761; Signaling System No. 7—ISDN User Part.

ITU-T Q.771; Functional Description of Transaction Capabilities.

ITU-T Q.772; Transaction Capabilities Information Element Definitions.

ITU-T Q.773; Transaction Capabilities Formats and Encoding.

ITU-T Q.774; Transaction Capabilities Procedures.

ITU-T Q.775; Guidelines for Using Transaction Capabilities.

ITU-T Q.920-Q.921; DSS1 Data Link Layer.

ITU-T Q.922; ISDN Data Link Layer Specification for Frame Mode Bearer.

ITU-T Q.931; ISDN User-Network Interface Layer 3 Specification for Basic Call Control.

ITU-T Q.932; Generic Procedures for the Control of ISDN Supplementary Services.

ITU-T X.25; Interface Between Data Terminal Equipment (DTE) and Data Circuit-Terminating Equipment (DCE) for Terminals Operating in the Packet Mode and Connected to the Public Data Networks for Dedicated Circuits.

ITU-T X.31; Support of Packet Mode Terminal Equipment by ISDN.

ITU-T X.200; Reference Model of Open Systems Interconnection.

ITU-T X.208; Specification of Abstract Syntax Notation One (ASN.1).

ITU-T X.209; Specification of Basic Encoding Rules for ASN.1.

ITU-T X.217; Service Definition for the Association Control Service Element.

ITU-T X.219; Remote Operations: Model, Notation, and Service Definition.

ISO 7498; Information Processing Systems—Open Systems Interconnection—Basic Reference Model.

ISO 8824; Information Processing Systems—Open Systems Interconnection—Specification of ASN.1.

ISO 8825; Information Processing Systems—Open Systems Interconnection—Specification of Basic Encoding Rules for ASN.1.
ISO 8649; Information Processing Systems—Open Systems Interconnection—Service Definition for the Association Control Service Element.
ISO 9072-1; Information Processing Systems—Text Communication—Remote Operations, Part 1: Model Notation and Service Definition.
ITU-T X.227; Association Control Protocol Specification for Open Systems Interconnection.
ITU-T X.229; Remote Operations: Protocol Specification.
ISO 8650; Information Processing Systems—Open Systems Interconnection—Protocol Specification for the Association Control Service Element.
ISO 9072-2; Information Processing Systems—Text Communication-Remote Operations, Part 2: Protocol Specification.
ISO 9545; Information Processing Systems—Open Systems Interconnection—Application Layer Structure.

Bellamy, John C., *Digital Telephony,* John Wiley & Sons, 1982.
Horrocks, R. J., and R. W. Scarr, *Future Trends in Telecommunications,* John Wiley & Sons, 1993.
Rose, Marshall T., *The Open Book. A Practical Perspective on OSI,* Prentice Hall, 1990.
Rumbaugh, James, et al, *Object-Oriented Modeling and Design,* Prentice Hall, 1991.
Smoot, Carl-Mitchell, and John S. Quarterman, *Practical Interworking with TCP/IP and UNIX,* Addison-Wesley, 1993.
Standards Committee T1—Telecommunications, *Procedures Manual,* Eighth Issue, 1993.
Steedman, Douglas, *ASN.1—The Tutorial & Reference,* Technology Appraisals, 1990.
Waters, Gill, *Computer Communication Networks,* McGraw-Hill, 1991.
Wilkie, George, *Object-Oriented Software Engineering,* Addison-Wesley, 1993.

Index

ABOUT THE AUTHOR

Arkady Grinberg is currently a member of Bellcore's technical staff involved in developing SCAI standards and network planning. He chaired a group in T1S1 that originally developed the SCAI standard. His 15-year career in the telecommunications industry—during which he has worked with Northern Telecom, Honeywell, Siemens, and Bellcore—has involved him in such activities as software design for switching systems, new switching system design, product management of packet switching and signaling equipment, network planning, personal communications services, and telecommunications standards development.